MEMS Reference Shelf

Series Editors:

Stephen D. Senturia
Professor of Electrical Engineering, Emeritus, Massachusetts Institute of Technology, Cambridge, Massachusetts

Roger T. Howe
Professor, Department of Electrical Engineering, Stanford University, Stanford, California

Antonio J. Ricco
Small Satellite Division, NASA Ames Research Center, Moffett Field, California

For further volumes:
http://www.springer.com/series/7724

Muthu B.J. Wijesundara • Robert G. Azevedo

Silicon Carbide Microsystems for Harsh Environments

Springer

Muthu B.J. Wijesundara
Automation & Robotics Research Institute
The University of Texas at Arlington
Arlington, Texas
USA
muthuw@uta.edu

Robert G. Azevedo
Proteus Biomedical Inc.
Redwood City, California
USA
razevedo@proteusbiomed.com

ISSN 1936-4407
ISBN 978-1-4614-2882-4 ISBN 978-1-4419-7121-0 (eBook)
DOI 10.1007/978-1-4419-7121-0
Springer New York Dordrecht Heidelberg London

Softcover reprint of the hardcover 1st edition 2011

Printed on acid-free paper

Springer is part of Springer Science+Business Media (www.springer.com)

This book is affectionately dedicated to my beloved parents for teaching me courage and perseverance and believing in me when others doubted.

-M. B. J. Wijesundara

To Miles, Ava, and Chloe.

-R. G. Azevedo

Foreword

When I was approached by my good friends, Drs. Muthu Wijesundara and Robert Azevedo, to write a forward for their book, I was truly honored. Each had worked, yes even truly struggled, with me to master some part of the technology of silicon carbide in our group efforts to bring silicon carbide forward as a complete RF wireless sensor technology. From the prospective developed in my service as DARPA program manager, Mechanical Engineering Department Chair, fellow researcher and Dean of the College of Engineering, I do attest that this wonderful book will be useful to the practitioner as well as to the researcher in the field of silicon carbide. Indeed, it is required reading now in my own research laboratory for all who would work with me in the research of silicon carbide. I have worked hard for over ten years in the struggle to master silicon carbide, and I am pleased to report to the reader that this volume will certainly ease your way toward a true command of the subject. I have reviewed the volume in great detail, and I affirm that it is useful both to the neophytes, eager to enter the field, as well as to the experts, seeking to deepen their knowledge. As a long-time member of the research community, with twenty-eight years of service to the University of California, Berkeley, I affirm that there is no equivalent book available today. And I offer my thanks and gratitude to my good friends for making all the effort to compile this edition — they have more diligence and fortitude than I, who have not yet compiled any book of my own.

With sincerity and admiration,

University of California, Berkeley, March 2011 *Albert P. Pisano, Professor*

Preface

This book is intended for the practicing microelectromechanical sensor designer as well as engineers and engineering managers in other fields. This book provides an introduction to harsh environment sensor applications and silicon carbide microelectronics and microelectromechanical system technology for such applications. Namely, this book reviews why silicon carbide is an excellent match for producing harsh environment microsystems, how silicon carbide substrates and films are produced and patterned, review progress towards silicon carbide microelectronics and microelectromechanical sensors, and how electronics and microsensors can be integrated and packaged. Various approaches to communication and power are also discussed. It is hoped that by providing a review of the pieces of silicon carbide microsystem technology currently available and outlining additional innovations needed to produce reliable harsh environment microsystems, new research will address these challenges and the full benefit of silicon carbide microsystems will be realized.

We came into silicon carbide technology through our research at the University of California at Berkeley. We worked together on materials and electromechanical design of harsh environment sensors made from silicon carbide for the DARPA HERMIT project. We would like to thank Prof. Albert P. Pisano and Prof. Roya Maboudian for introducing us to this research.

Others have contributed to this manuscript, so we would like to offer our sincere thanks. Dr. David Myers, Dr. Anand Jog, and Raminderdeep Sidhu reviewed drafts of the manuscript and provided valuable input into the structure and flow. Steven Elliot facilitated various logistics and feedback throughout this process.

We would also like to give a special thank you to our families, who have been understanding of nights or weekends spent working on this book and whose support is, as always, what enables us to pursue our passions.

Arlington, Texas — *Muthu B. J. Wijesundara*
Albany, California — *Robert G. Azevedo*
March 2011

Contents

1 **Introduction** ... 1
1.1 Harsh environment applications ... 2
1.1.1 Combustion monitoring ... 3
1.1.2 Process control ... 6
1.1.3 Military and Space Navigation ... 8
1.1.4 Structural Health Monitoring ... 9
1.1.5 Space Exploration ... 14
1.2 SiC properties ... 15
1.2.1 SiC crystal structure ... 15
1.2.2 Basic Chemical and Physical Properties ... 16
1.2.3 Electrical ... 18
1.2.4 Thermo-mechanical ... 24
1.3 System integration aspects of SiC materials ... 26
References ... 29

2 **SiC Materials and Processing Technology** ... 33
2.1 Material considerations for various applications ... 33
2.1.1 Electronics (crystallinity, doping, defects, polytype) ... 33
2.1.2 MEMS (Stress, strain gradient) ... 35
2.2 Substrate ... 36
2.2.1 Vapor Phase Growth ... 37
2.2.2 Solution Phase Growth ... 44
2.2.3 Growth Related Issues ... 46
2.2.4 Current Status ... 51
2.3 Epitaxial Thin Films ... 53
2.3.1 Vapor Phase Epitaxy (VPE) ... 53
2.3.2 Liquid-Phase Epitaxy (LPE) ... 64
2.3.3 Vapor-Liquid-Solid Epitaxy (VLS) ... 64
2.3.4 Current Status of SiC Epitaxial Wafers ... 65
2.4 Polycrystalline SiC and Amorphous SiC Films ... 66
2.4.1 APCVD ... 67

2.4.2 LPCVD ... 67
2.4.3 PECVD ... 74
2.4.4 Ion Beam Assisted Deposition ... 75
2.4.5 Other Deposition Methods ... 78
2.5 Patterning of SiC ... 80
2.5.1 Dry Etching of SiC ... 80
2.5.2 Wet Etching of SiC ... 82
2.5.3 Molding ... 83
2.5.4 Other Patterning Methods for SiC ... 84
2.6 Planarization and Surface Preparation ... 85
2.6.1 Lapping and Mechanical Polishing ... 85
2.6.2 Chemomechanical Polishing ... 86
2.6.3 Surface Preparation ... 87
References ... 87

3 Silicon Carbide Electronics ... 97
3.1 General semiconductor process flow ... 97
3.2 Ion implantation doping ... 100
3.3 Contacts to SiC ... 102
3.3.1 Ohmic contacts ... 103
3.3.2 Schottky contacts ... 105
3.4 Electronics for high power applications ... 106
3.4.1 Schottky Rectifier ... 106
3.4.2 JBS Rectifier ... 109
3.4.3 TSBS Rectifier ... 111
3.4.4 P-i-N Rectifier ... 111
3.4.5 MPS Rectifier ... 112
3.4.6 BJT ... 114
3.4.7 SIAFET ... 116
3.4.8 Power MOSFET ... 116
3.5 Sensor interface electronics for high temperature applications ... 117
3.5.1 MOSFET and MESFET considerations ... 118
3.5.2 JFET ... 119
3.5.3 BJT ... 120
3.6 Electronics for communication electronics ... 121
References ... 122

4 SiC MEMS devices ... 125
4.1 SiC MEMS processing ... 125
4.1.1 Surface (thin-film) micromachining ... 126
4.1.2 Bulk micromachining ... 129
4.2 Devices and operation in harsh environment ... 132
4.2.1 Strain ... 132
4.2.2 Pressure ... 138
4.2.3 Inertial Sensing ... 144

4.2.4 Chemical ... 149
4.2.5 Photo and Radiation Detectors ... 154
4.2.6 Temperature ... 154
4.3 Material aspect of harsh environment operation ... 155
4.3.1 CTE of layers and effect ... 156
4.3.2 Stress gradient and stress relaxation ... 159
4.3.3 Design for high g-shock survivability ... 160
References ... 163

5 Packaging ... 167
5.1 Zero-level MEMS Packaging ... 168
5.1.1 Bonding Techniques ... 170
5.1.2 Deposition Techniques ... 173
5.2 First-level MEMS Packaging ... 182
5.2.1 Thermal Expansion Matching of First-level Packaging Materials ... 183
5.2.2 Electrical Interconnects ... 184
References ... 186

6 System Integration ... 189
6.1 SiC MEMS and SiC Electronics Integration ... 189
6.1.1 Multi-chip Hybrid Approach ... 190
6.1.2 Monolithic Fabrication Approach ... 191
6.2 Power ... 200
6.2.1 Standalone Power Sources ... 201
6.2.2 Environmental Power Scavengers ... 205
6.2.3 Remote Powering (RF powering) ... 215
6.3 Communication ... 217
6.4 Testing of Harsh Environment Electronics, MEMS and Microsystems ... 217
6.4.1 High Temperature Testing of SiC Electronics ... 218
6.4.2 High Temperature Testing for SiC MEMS ... 218
6.4.3 Shock Survivability Testing ... 220
6.5 Gallium Nitride, A Prospective Material for Harsh Environment Microsystems ... 221
6.6 Current state and Future Prospective ... 224
References ... 226

Index ... 231

Chapter 1
Introduction

Microsystems, or microelectromechanical systems (MEMS), technology continues to grow rapidly by enabling ever emerging applications that demand diverse, versatile functionality. Microsystems refers to a class of sub-millimeter scale sensors and actuators coupled with signal processing capable of measuring physical and chemical changes or performing desired physical and chemical functions. Microsystem technology based on micro-scale mechanical transducers progressed because silicon (Si) possesses both favorable electrical and mechanical properties to create these micro-sensor elements. Although many types of materials, ranging from ceramics to polymers, have been explored as platforms for microsystem technology, Si is currently the dominant platform. Si microsystems leverage the highly-parallel batch fabrication paradigm that has made microfabricated silicon-based semiconductor electronics commercially viable. Furthermore, they have benefited from a large body of knowledge around Si masking and etching techniques, which make fabrication of complicated geometries possible. This has enabled the current pervasiveness of silicon microsystems and components; they range from accelerometers for automotive airbags and inertial sensing, gyroscopes in video game controllers, micro-mirrors for projection displays, injector nozzles for inkjet printer cartridges, and mechanical timing references and RF filters for communication systems.

Diverse functionality, enhanced performance, small form-factor, and batch fabrication capability are the primary driving forces to using micro-scale systems technology. Smaller size makes it possible for noninvasive integration to existing systems. Batch fabrication promises more functionality at competitive price points compared to more traditional manufacturing techniques. Enhanced performance comes from two aspects of microsystems. A more direct integration of the sense elements with their electronics reduces noise. Sensitivity is increased by being able to make delicate structures that are inherently more sensitive or by utilizing physical phenomenon that dominate the response of these micro-structures but not their macro-scale counterparts. Given the considerable success to date of silicon microsystems for improving performance of a variety of commercial sensors, research has looked to expand the application space into harsh environments, such as combustion control and chemical process monitoring [1]. This has elevated the need for being able to use material systems compatible with harsh enviroment conditions to create sensors that can be used in an even wider range of applications than silicon.

Silicon carbide (SiC) is a leading alternative to Si in any application in which the environmental temperatures or level of radiation would damage silicon electronics or temperatures and corrosion would damage silicon or polymer sensor elements. SiC, like silicon, is usable as a semiconductor electrical element. In addition, it possesses mechanical behavior that is comparable to silicon. At high temperature, silicon dramatically softens and dopants diffuse, changing electrical behavior. Conversely, SiC does not significantly soften even at 600 °C and has a very low diffusivity. SiC is also highly resistive to a wide range of chemicals and can withstand high-temperature oxidative environments. Despite the chemical resistance of SiC, silicon-compatible etching and polishing processes have been developed for SiC. This allows a similar degree of flexibility in layout as silicon or polymer MEMS. Because of its overall stability of mechanical and electrical properties at high temperatures along with a developed batch manufacturing infrastructure, SiC is poised to open up a host of harsh environment applications to microsystem technology [2, 3, 4, 5].

This book aims to put into context how SiC technology can play a role in enabling sensors for a wide range of harsh environment applications not currently served by Si technology. It will also provide an updated view of SiC technology from deposition techniques, electronic device and microstructure fabrication, and packaging. It will conclude with a discussion of challenges in putting these parts together into a complete harsh environment microsystem, highlighting important research gaps in order to help spur interest in these areas. It aims to provide a useful reference for the MEMS practitioner yet remain accessible to those who are new to the field of microsystem technology.

To set the context for the remainder of the book, the remainder of this chapter will outline several harsh environment application spaces before outlining what makes SiC the right material system for developing microsystem technology for these applications.

1.1 Harsh environment applications

Harsh environment conditions are identified as undergoing large temperature excursions or required to operate for extended time at high temperature, exposure to highly corrosive and erosive conditions, intense radiation exposure, and high shock or intense vibration. These conditions are common to combustion environments, certain forms of energy and chemical production, space exploration, military-grade vehicle control sensing, and munitions monitoring (Figure 1.1). Microsystems capable of operating in the above environments with a high degree of reliability, efficiency, and sustainability are categorized as harsh environment microsystems. To extend the existing Si- or polymer-based commercial technology and techniques to harsh environment applications, development of robust material systems and device

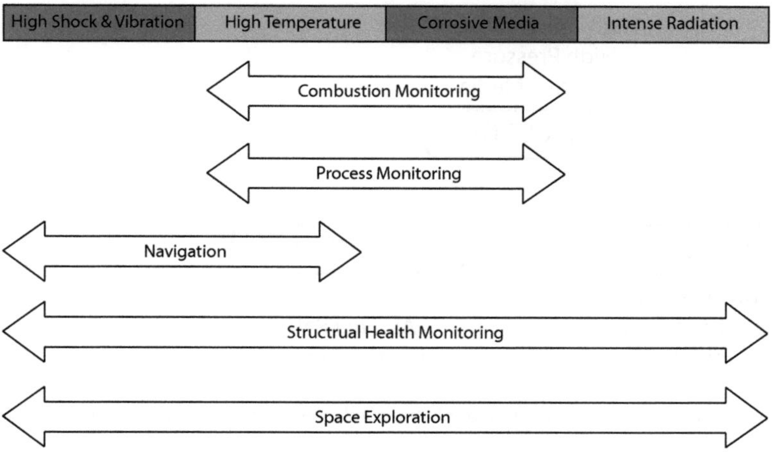

Fig. 1.1 Harsh environment conditions relevant to various harsh environment applications.

technology is needed. Requirements and benefits for each harsh environment application area are detailed next.

1.1.1 Combustion monitoring

Stringent environmental regulation and the demand for higher fuel economy require ever-increasing advances in combustion controls for automotive engines and gas turbines [6]. Combustion efficiency of automotive engines and gas turbines vary due to manufacturing discrepancy, aging of engine components, fuel properties, and intake air qualities (humidity and temperature) [7, 8]. For optimal efficiency and emission, precise control of temperature, pressure, air-to-fuel ratio, and ignition timing is required [9]. Traditional sensing methods are generally indirect with respect to the combustion event. Although they have been effective in further reducing emissions and improving fuel economy, further design optimization of these systems is seen to have limited additional improvements in emissions control and fuel efficiency. Therefore, research has shifted focus to active sensing and control of the combustion event in order to meet future regulation limits on emissions and higher fuel efficiency standards [8, 12].

Figure 1.2 schematically represents the overall concept of direct combustion monitoring. Increasing efficiency, lowering emission, and adapting fuel flexible technologies are key benefits that can be achieved through real-time monitoring and control of combustion events. Stoichiometric fuel burning and control of combustion temperature are required for reduction of CO, NO_x, and hydrocarbons. Figure 1.3 shows how precisely the air-to-fuel ratio has to be controlled to achieve maximum efficiency and low emissions. On-board real-time monitoring of the combustion event allows optimization of combustion based on current fuel properties, air

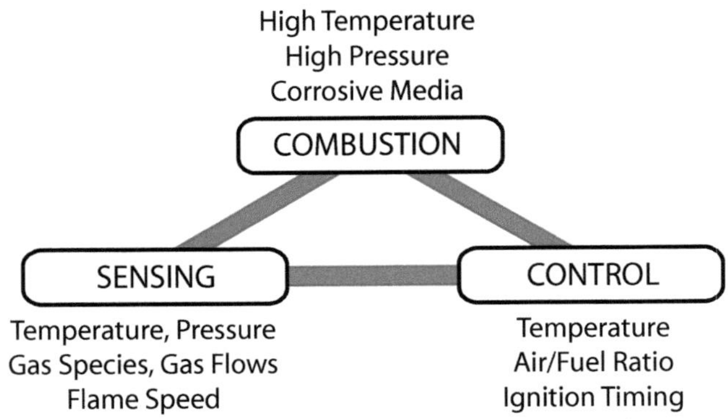

Fig. 1.2 Schematic representation of active closed loop combustion control through real time monitoring.

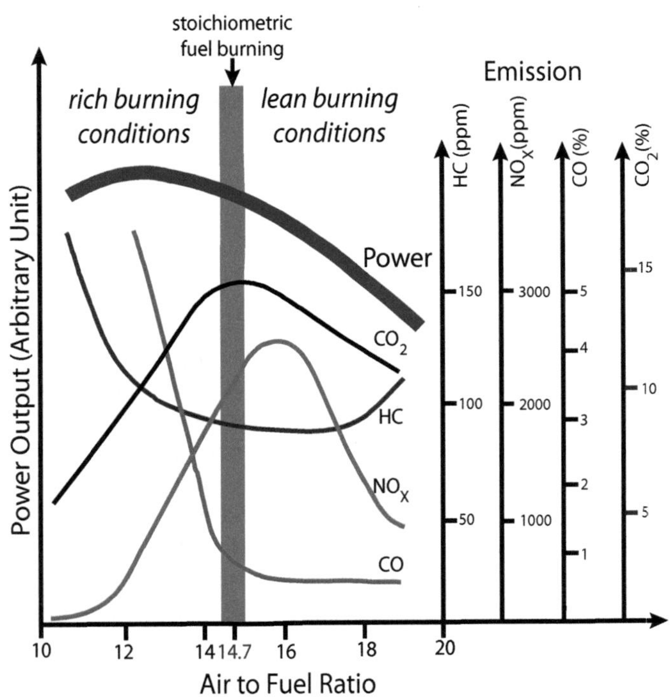

Fig. 1.3 Emission and power output of typical gasoline automotive engine with respect to air to fuel ratio. (Data from [9], [10], and [11] used to create the plot.)

quality, and engine conditions. Active control allows optimization of the combustion process at all times and under differing load conditions so that the ideal range for temperature, pressure, and fuel-to-air ratio can be maintained. This kind of system will also enable fuel flexible operation as the engine can optimize the combustion

Table 1.1 [1] Typical conditions and anticipated time of operation of combustion sensors for automotive and gas turbine applications (©Elsevier 1999), reprinted with permission.

Application	Temperature [°C]	Conditions	Time of Operation [hrs]
Gasoline engine	200-450	pressure, oxidative	4000
Diesel engine	200-450	pressure, oxidative	6000
Gas turbine	400-650	pressure, vibration, oxidative	10000

parameters suited for the particular fuel type regardless of slight variations in the fuel constituent.

In particular to automotive engines, combustion efficiency varies cylinder to cylinder, compromising the overall efficiency of the engine and increasing emissions [13]. This variation is attributed to factors such as injector and air intake variability and temperature variation. Thus, individual cylinder control has become a topic of major interest for engine development. One of the approaches widely discussed is the use of in-cylinder combustion monitoring with closed-loop combustion control. Because of the need to be as close to the combustion event as possible to achieve the highest benefit from closed-loop control, harsh environment sensors will play a critical role in achieving this task.

Gas turbine optimization for high efficiency and reduced emissions focuses on combustion instability and burn pattern. Active control with embedded feedback sensors that perform dynamic pressure and air-to-fuel mixture measurements as well as temperature mapping are needed to directly dampen unwanted combustion instabilities and optimize the burn pattern. Microsensors that can operate in combustion conditions permit incorporating these systems into combustion chambers without an adverse effect to turbine operation.

In order to achieve the forecasted need for improved efficiency and reduced emissions, sensors that can measure combustion parameters such as temperature, pressure, flame speed, and chemical species are needed. MEMS-based sensors are of considerable interest because their small form factor potentially allows them to be directly incorporated into the combustion environment with negligible perturbation to the combustion event and to the chamber infrastructure. However, these sensors must withstand combustion conditions such as temperature, mechanical loads (pressure, vibration), and chemically aggressive media while still being able to perform reliably for an extended period of time. Table 1.1 summarizes the typical conditions and anticipated time of operation of combustion sensors for automotive and gas turbine applications. As will be detailed later, these requirements exceed the limits of silicon-based microsensors and microsystems. Thus, alternative, high temperature materials are required.

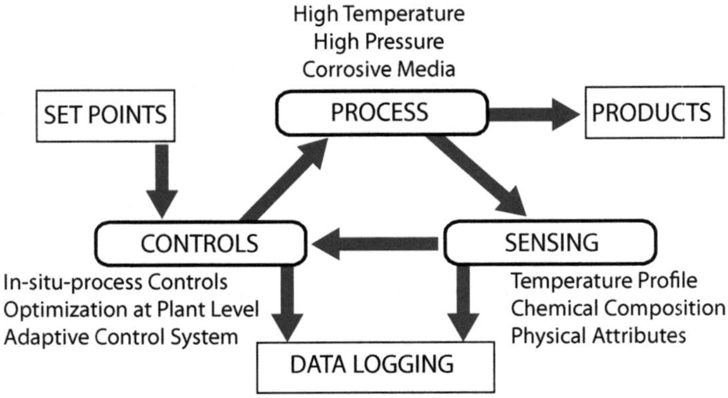

Fig. 1.4 Schematic flow of how sensor data can be used in industrial process control.

1.1.2 Process control

In any production environment, increasing production, maintaining quality, reducing operational cost, and minimizing waste are the common goals. Process quality monitoring through periodic sampling and product quality assessment leads to a lag before detecting out-of-tolerance conditions, which in turn results in increased costs due to production of inferior product that typically must be disposed of as well as wasted manufacturing resources. As with combustion monitoring, *in-situ* real-time monitoring of process parameters, process conditions, and product quality will allow active control that achieves the highest efficiency and standards because it is the most direct measurement of the system to be controlled [14, 15]. It also provides data logging capability for long-term planning as well as process and equipment improvements (Figure 1.4). Many industrial processes in energy and chemical production are extremely harsh. Sensors must survive these conditions for a extended period of time to be utilized in active control. If the sensors can also be made compact enough, it would be possible for direct implementation inside critical reactors and ancillary equipment. Microsystems can achieve the desired form factor. So harsh environment microsystem research is targeting improved chemical robustness to realize active control for these types of applications. Two major industries that can vastly benefit from harsh environment microsystem technology are oil exploration and extraction and chemical production.

1.1.2.1 Oil and Energy

Oil and gas extraction has changed tremendously as the demand for fossil fuels increases. For instance, petroleum wells have changed from simple vertical wells to more complex subsurface networks comprising of horizontal, multi-lateral, and multi-branched wells (Figure 1.5). This complexity has introduced new challenges

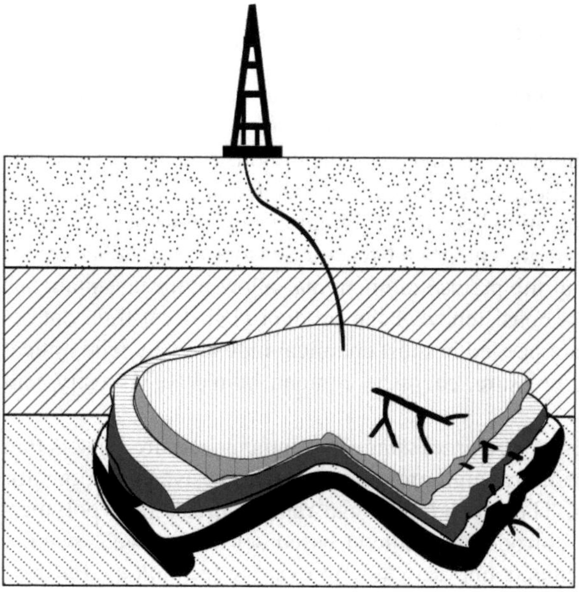

Fig. 1.5 Schematic representation of a multi-lateral, multi-branched oil well.

in reservoir management in order to increase oil and gas production, maximize recovery, and at the same time minimize operation cost.

The current practice of gathering down-hole information through periodic interrogation and data logging is costly and has a risk of damaging the well. Additionally, optimization of oil production is rather difficult. Down-hole sensors are still typically in the proposal or research stage. However, there are some limited deployments, such as the use of fiber Bragg grating sensors, to acquire dynamic and static conditions of the well [16]. Due to the complexity of these methods, there are ongoing efforts to develop better technology for distributed sensors throughout the well to get real time data on temperature, pressure, flow rates, and water content as well as distributed actuators to control fluid flow. The data from these sensors along with flow control systems will allow real-time optimization and improve tactical reservoir management and strategic planning. But the down-hole environment requires survival in liquid and often corrosive media as well as exposure to high pressures and temperature [17]. Given recent concerns with deep water drilling, distributed sensing may also provide another layer of safety to these types of installations. If these sensor and actuator modules can be controlled wirelessly, wide spread usage is expected; however, the distances and media involved are difficult challenges for reliable wireless communication.

1.1.2.2 Chemical Manufacturing

The traditional product control methodology often leads to sub-optimal equipment utilization, long delay time when transitioning product grades, and changes in product qualities at start-up and shutdown in many chemical industries. In part, the difficulty lies in indirect measurement techniques and associated lag time between the chemical process and sensor location because of limitations in sensor survivability.

Similar to oil well applications, integrated reactor sensors and feedback control units can detect changes at the precise moment and allow more instantaneous remedies, which will improve consistency of the product [18]. Real-time data is the key. Typical sensors that are needed in most chemical industries include optical properties, temperature, flow, pressure, chemical content, humidity, viscosity, and particulate and slag detectors.

In most chemical industries, severe processing environments require sensors that operate reliably at high temperature, high pressure and in corrosive conditions for active control of the process. These conditions can rapidly breakdown polymer-based sensors and can often corrode even silicon-based sensors as well, making certain direct reactor measurements impractical with existing technology. Furthermore, the operating temperatures may lead to premature failure or outright exceed the operating limits of silicon electronics, requiring remote sensor signal processing with respect to the reactor environment.

1.1.3 Military and Space Navigation

High precision navigation is a prerequisite for most military and aerospace launch vehicles. Real-time three-dimensional position data and precise time measurement along with thrust and trajectory controls are essential for successful mission completion [19]. This is true for both manned and intelligent unmanned operations including space shuttles, rockets, missiles, UAV, and even planetary probes like the Mars rover. Current inertial measurement units (IMUs), which incorporate gyroscopes and accelerometers with sense electronics to provide information on up to six degrees of freedom, work in conjunction with position control thrusters to provide the necessary functionality. The current units are large, heavy, and require a significantly large power source as well as bulky cooling systems. This is primarily due to high temperature and extreme shock and vibration environment typical in military and space applications. In most military and space applications, launch or landing impact can exceed 100,000 g; thus high g survivability is key [20, 21]. Space exploration missions additionally subject sensors to high levels of radiation and in some cases corrosive gases.

Silicon-based MEMS position sensing systems are being fabricated in an integrated fashion making these systems extremely small and low weight. Gyroscopes, accelerometers, and control electronics integrated into small IMUs not only decrease size and weight but also tend to reduce power consumption. This allows these systems to be applied to new and existing systems with minimal intrusion to the infrastructure. Low power, smaller size, and reduced weight also allow applications in lighter, faster, and smaller-sized launch vehicles or aircraft critical for future combat and aerospace operations.

State-of-art silicon-based technology is not compatible with the operating conditions encountered by most of these systems without supporting infrastructure such as cooling, extensive packaging, and radiation shielding. These additional items add more volume and weight, reducing the inherent size and weight advantage of MEMS-based systems. Especially in the demanding field of space exploration, but even for more traditional military applications, all of these weight and size factors must be minimized in order to decrease the overall cost of the mission.

Hence, more robust systems that can also survive the application environment with a dramatically reduced supporting infrastructure are needed to maximize the benefit of payload reduction by adopting microsystem technology. If instead high temperature and shock resistant inertial navigation microsystems are developed that can in some cases be used directly, with no external cooling or sophisticated packaging, that in turn reduces weight, decreases complexity, and enhances reliability.

1.1.4 Structural Health Monitoring

All engineered structures undergo material and geometrical changes over time due to use, environmental effects, and manufacturing anomalies. The process of measuring these structural changes is referred to as structural health monitoring (SHM) [22]. The most common monitoring process involves periodic evaluation of structural properties; however, periodic evaluation provides limited knowledge. In the worst case, this can lead to unsafe conditions and loss of life. It also leads to unscheduled downtime and often extensive labor costs due to the unscheduled nature of the event. Most importantly, unanticipated failures can occur between inspection intervals. The real-time monitoring of in-service structures with less human intervention mitigates the shortcoming of periodic-inspection-based SHM [23].

A variety of sensors capable of continuous monitoring of structural response are commercially available. These are commonly piezoresistive, piezoelectric, or fiber-optic based. Real-time monitoring requires integrating these sensors and signal processing units into critical areas of the structure. The integration and networking complexity depends on sensor type, structure type, and its dynamic state. Additionally, the physical complexity of a structure as well as periodic maintenance, which may involve replacement of structural components, make wire based solutions less attractive or prohibitively costly. Combining SHM sensor nodes wirelessly is an important aspect of the system design.

The ideal wireless sensor network module for SHM should consume very little power, be small, be reliable, and be stable over long intervals, require no real maintenance, and be capable of operating in a wide array of environments. MEMS-based systems are drawing considerable attention due to their versatile functionality,

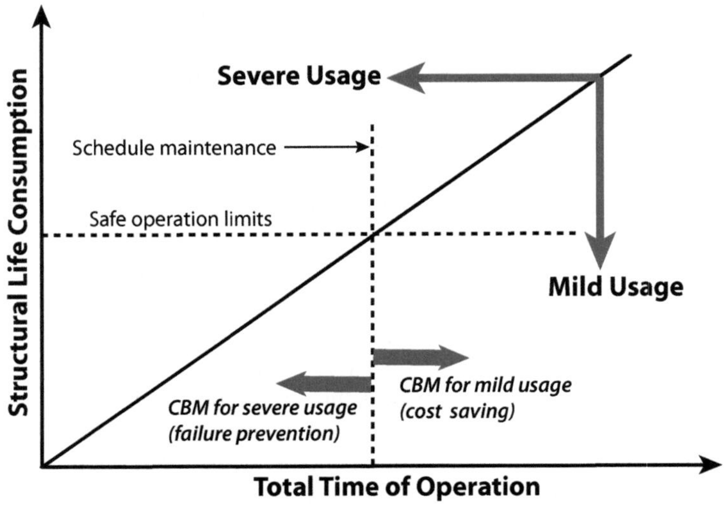

Fig. 1.6 Usage-based maintenance for failure prevention as well as for cost reduction. (Figure created using the concept presented in [24].)

low power requirement, small form factor, and high sensitivity. A small form factor sensor system with high sensitivity is very attractive since it can be installed onto many complex structures without the need to compromise the design of the original structure.

With continuous SHM, maintenance activities can be scheduled based upon actual conditions or usage of the system. This is called condition-based maintenance (CBM). Manufacturer recommended maintenance intervals can be either too aggressive based on actual use, leading to unnecessary costs, or too lenient in some circumstances, which can result in preventable failures. This generally becomes an issue because the manufacturer recommended intervals do not capture all the relevant operating conditions. For instance, sometimes early maintenance is necessary when heavy usage or environmental causes accelerate structural degradation. Or in-service structures can experience premature failure due to manufacturing imperfections, either in design or materials. Figure 1.6 graphically represents the advantage of usage-based maintenance for failure prevention as well as for cost reduction.

An extension of CBM is adaptive operation. The use of dynamic and static responses of in-service structures to optimize real-time operating conditions is called condition based adaptive operations (CBAO). As structures age, the likelihood of failure increases. Based on real-time structural health data, better operational conditions can be found to ensure safety and performance before maintenance is done. In this case, CBAO allows for safe extension of the maintenance interval when immediate down-time for the structure would prevent critical operations or interrupt sensitive missions. This can also extend to operating with known damage or under unusual external disturbances, in which case regular operational parameters have to

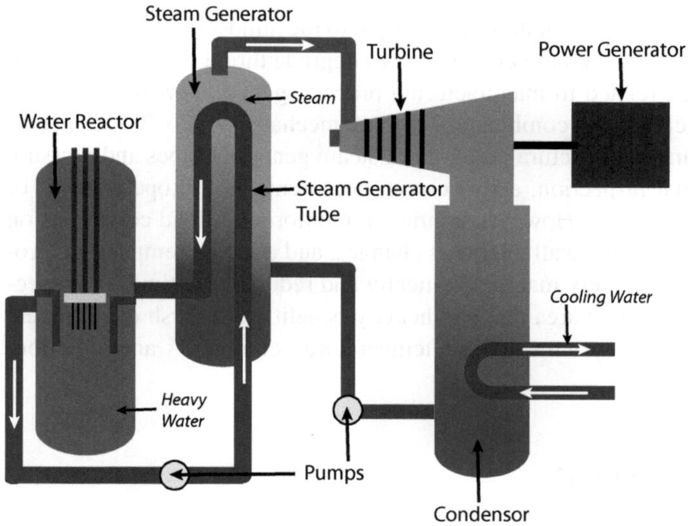

Fig. 1.7 Generalized architecture of a nuclear power plant.

be drastically adjusted to ensure safety until proper maintenance can be performed or the disturbance passes. In some situations, such as high-value military missions, it is desired to allow systems to push beyond their normal operating limits for limited intervals without sacrificing safety. For instance, monitoring structurally critical components for in-use deformation signatures can allow periodic relaxation of standard speed limitations during critical missions since real-time monitoring can facilitate development of less conservative safety factors. CBAO algorithms can use this information to both allow for overload conditions and adjust subsequent maintenance schedules for these equipment accordingly.

If these MEMS-based sensor systems were additionally capable of working in harsh environment conditions, these benefits could be extended from current work on bridges and buildings to specific critical needs in energy, avionics and aerospace, shipping, and chemical industries. Details of how harsh environment SHM would benefit these areas are discussed next.

1.1.4.1 SHM for Nuclear Energy Production

Figure 1.7 represents a generalized nuclear power plant. Pressure vessel and steam generator tubes are critical parts of a nuclear power reactor. Pressure vessels surround the nuclear reactor core that produces super-heated water. The steam generator tube exchanges this heated water from the core reactor side to the steam generation side for driving turbine generators. The structural integrity of both the pressure vessel and steam generator tube is vital as they contain water with radioactive fission products. Specifically, any leaks in the steam generator tubes could lead to the escape of nuclear fission materials directly into the atmosphere in the form of steam. Pressure vessel and steam generator tubes degrade through many paths that include tube damage related to manufacturing process, general corrosion, pitting, stress corrosion cracking, or a combination of these mechanisms [26, 27]. Common methods of inspecting the structural integrity of steam generator tubes and pressure vessels include visual inspection, eddy current measurements, isotope analysis of steam, and helium leak tests. However, continual monitoring of load conditions on vulnerable areas, monitoring wall thickness changes, and tracking temperature profiles would increase the safely margin drastically and reduce downtime for inspections and repair. This is one area that can heavily benefit from harsh environment microsystems, especially systems that are temperature-, corrosion-, and radiation-hard.

1.1.4.2 SHM for Naval Vessels

Monitoring the hull structure is vital for any naval operation as most failures occur due to compromised hull integrity. Detection of hull thickness reduction rates, identifying localized corrosion, and tracking fatigue accumulation are highly critical for minimizing risk factors, optimizing operation, and improving ship fleet management [28]. Aging of naval vessels highly depends on operating conditions. For instance, vessels that operate in rough seas experience a higher degree of stress and fatigue compared to those in calm waters. Thus, continuous monitoring and data logging is important for maintenance scheduling. Real-time diagnosis of the ship hull response to rolling motion, wave slamming, and acceleration caused by ship movements and sea states allows adaptive control of vessel maintenance [29].

Sensors that are suited for naval structure monitoring include strain gauges for measuring structural flexure, pressure transducers for measuring emergence and slam, accelerometers for measuring vertical motion, and inclinometers to measure pitch and roll [30, 31]. Wireless sensors systems are ideal for naval vessels as many of the ship structures undergo frequently scheduled maintenance and parts replacement. MEMS are particularly attractive due to their small size, high sensitivity, and integration capability, which allows minimal effects on the structure and its functionality. However, the micro sensor systems applied to naval systems need to also withstand large temperature fluctuations and corrosive atmosphere present in naval environments.

1.1.4.3 SHM for Aerospace

The fuselage of an aircraft or spacecraft must withstand the loads anticipated during the service life of the aircraft regardless of the age [23]. Corrosion, erosion, stress-induced fatigue, and accidental damage are common causes of structural failure in airplane and space shuttle structures. Traditionally, inspections and maintenance are done at regularly scheduled intervals or when a problem is identified. These ap-

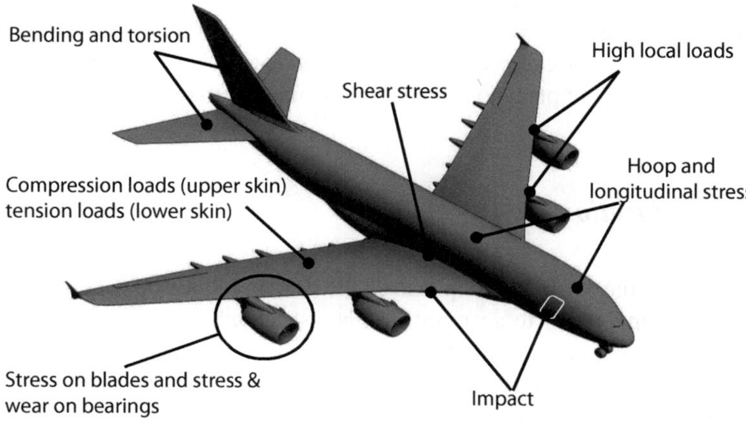

Fig. 1.8 [34] Loading points of aircraft structures (©ASME 2006), reprinted with permission.

proaches consume substantial resources and time. Furthermore, a catastrophic failure can occur, similar to that of Aloha 737, despite the above procedures, caused by a particular set of operational conditions not adequately captured by standard maintenance scheduling [33]. Aging of an aircraft component cannot be determined only by flying time or number of flights. It is also a function of operational conditions such as hard landings, severe turbulence, rough runways, and short distance flying. Real-time diagnostic and prognostic systems can vastly improve safety and reliability. As depicted in Figure 1.8, jet engine turbine components, wing structures, leading edges, landing gear, fuel tanks, and the fuselage are some of the critical component that can benefit from continual structural monitoring [34].

Even though some accelerometer and strain sensor solutions are proposed or currently being used, MEMS-based sensor platforms are desired due to small form-factor and low weight. Adapting harsh environment compatible sensor modules can further reduce weight by reducing cooling requirements or by allowing sensors to be integrated into critical high-temperature components like the aircraft engine. To measure the structural health of certain critical moving components such as turbine blades and bearings in the aircraft engine, it is imperative that these sensor modules be wireless [35].

Structural degradation of most space vehicles start from the moment of launch [36, 37]. For example, thermal protection shield damage of the space shuttle can occur during lift-off. Thus, space launch vehicles are prime candidates for SHM. As is the case for naval structures, the following sensors would provide much needed structural data on aerospace structures: strain, vibration (accelerometers), displacement, and temperature. Outer space opens up another aspect of harsh environment survivability because of the increased exposure to radiation and even larger temperature excursions. At the extreme end, the gamut of conditions are encountered in NASA's Venus mission, which requires devices to operate around 500 °C at high-pressure conditions on the order of 90 atmospheres in the presence of carbon dioxide, sulfuric fumes, as well as chlorine and fluorine compounds. Silicon-based sensors and electronics are not expected to be suitable for such a mission without considerable infrastructure in place to protect them from the environment.

1.1.5 Space Exploration

Every space agency in the world has emphasized the need to reduce mission cost for future missions. The main way to accomplish this task is reduce the launch costs, which can be achieved simply by reducing the launch mass. For instance, reducing the launch mass of an interplanetary mission from 7800 kg to 750 kg will save nearly half a billion dollars just in the reduced cost of the launch rocket [38].

MEMS-based ultra-miniaturized systems for space applications have drawn considerable attention as a cost-saving measure by their considerable reduction in weight while being able to add new capabilities [40, 41, 42]. Low power requirements of these miniature systems also indirectly reduces mission weight by reducing the size of battery packs for instance. Currently, MEMS sensors are in early research or limited-use trials for many space expeditions. MEMS-based devices used in space missions include accelerometers, gyroscopes, pressure sensors, atomic force microscopes, and low-noise timing references for communications systems.

Even though current microsystems technology possesses the required maturity and reliability for many space applications, the limitations in intrinsic material properties hinder generalized application to many space environments. The systems applied into space applications should, as mentioned for SHM, have immunity to environmental conditions such as high temperature, high levels of radiation, and corrosive media [43, 44]. For current state-of-the-art microsystems technology, the cooling and heavy shielding infrastructure that is needed for environmental protection compromises the weight and size advantage over more traditional technologies. Thus, to fully tap into the weight and size advantages of microsystems, they should withstand the demanding conditions of space with little to no additional shielding or cooling components.

Some of the potential applications of harsh environment microsystems for space exploration have been discussed in previous sections such as navigation and structural health monitoring. At a component level, MEMS-based on-chip IMUs, RF switches and timing references, micro-valves, and micro-thrusters will enable new space mission scenarios. Applications such as micro-probes, micro-rovers, and aerobots will all benefit from the development of harsh environment microsystems technology. Furthermore, in addition to the weight-saving advantage of future manned and unmanned missions, new concepts like micro- and nano-satellite deployments will most likely depend on developing robust, reliable harsh environment microsystems that can meet the unique demands of space environments.

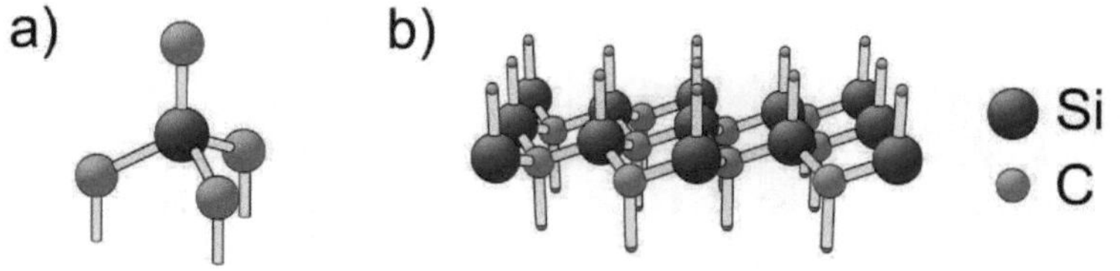

Fig. 1.9 [45] Atomic structure of SiC: a) tetrahedrally bonded Si-C cluster b) Hexagonal bilayer with Si and C in alternating tetrahedrally coordinated sites (©Springer 2004), reprinted with permission.

1.2 SiC properties

For the harsh environment applications outlined in the previous section, all the components of the microsystem must withstand either high temperature, high radiation, intense vibration, high G-shock, corrosive environments, or some combination. Thus, materials with robust chemical, electrical, and mechanical properties are needed. Silicon carbide (SiC) has been identified as the best suited material because of its unique materials properties. It is mechanically robust, chemically inert, and can be used as a semiconductor substrate for integrated circuits. Furthermore, SiC can be used to fabricate each and every component of the microsystem, namely electronics, sensors, and packaging.

1.2.1 SiC crystal structure

SiC can be found or produced in three forms: single crystalline, poly-crystalline, and amorphous. Depending on the device type and the functionality, all these forms can be useful in creating SiC microsystems. SiC substrates and electronics-grade epitaxial SiC are single-crystalline while most MEMS structures are fabricated using poly-crystalline SiC (poly-SiC). Amorphous SiC is useful for MEMS structures and isolation layers as well as device encapsulation.

Single-crystalline SiC exists in many different polytypes. However, most research has focused on just three types: 6H-SiC, 4H-SiC, and 3C-SiC (also known as β-SiC). In all polytypes, each silicon atom is bonded to four neighboring carbon atoms. Each carbon atom in turn bonds to four neighboring silicon atoms in a tetrahedral fashion (Figure 1.9a). These Si-C units are arranged in a hexagonal bilayer with Si and C alternately occupying sub-layers (Figure 1.9b). The stacking sequence of the Si-C bilayer with respect to the orientation of adjacent layer determines the polytype. 3C-SiC (cubic) is formed with an identical orientation of each bilayer and the atomic geometry is repeated every three layers along the c-axis of the crystal (Figure 1.10a). 4H-SiC (hexagonal) is formed by stacking blocks of two identically oriented bilayers to form the unit cell, but subsequent cells are rotated 60° with re-

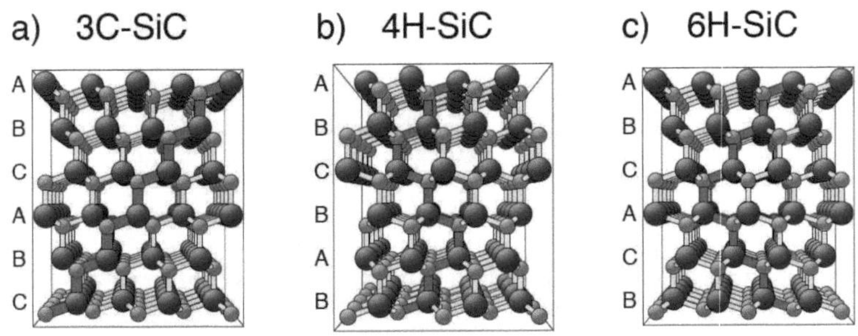

Fig. 1.10 [45] Crystal structure of different SiC polytypes displayed parallel to the $(11\bar{2}0)$ plane: (a) cubic 3C-SiC (β-SiC), (b) hexagonal 4H-SiC, and (c) hexagonal 6H-SiC (©Springer 2004), reprinted with permission.

spect to each other (Figure 1.10b). 6H-SiC (hexagonal) is formed with slabs of three identically oriented bilayers to form the unit cell. Again, the unit cell is rotated 60° with respect to the neighboring cell layers (Figure 1.10c).

Polycrystalline SiC, as the name implies, contains local regions of crystallinity; however, the crystals are not continuous throughout the layer. Adjacent islands of crystals may have different densities and crystal basal plane orientations. Poly-SiC is mostly produced as a thin film for MEMS device fabrication because of the reduced deposition temperatures required for poly-SiC fabrication over single-crystalline SiC. 3C-SiC is the most common polycrystalline polytype produced and can be deposited on various substrates including silicon and silicon carbide [46, 47]. Columnar or grain-type microstructure can be obtained by varying the deposition parameters. In addition to Si and C, poly-SiC may contain hydrogen or other residual elements depending the deposition methods.

Amorphous SiC is mainly use as a structural material for MEMS and encapsulation. The density and stoichiometry of the amorphous SiC are strongly dependent on deposition method and deposition parameters. Again, due to the lower deposition temperatures, residual elements such hydrogen and argon remain trapped in the film. Because of the lack of crystallinity, amorphous SiC acts as a dielectric.

1.2.2 Basic Chemical and Physical Properties

In this section, general physical and chemical properties are discussed. Electrical and mechanical properties are presented in following sections. SiC is a high temperature ceramic material. It sublimates at 2830 °C, which is very high compared to the melting point of silicon (1420 °C). The high temperature stability in the solid state is in part due to the extremely low diffusion rate in SiC. At high temperatures, most semiconductors undergo changes due to diffusion; however, significant diffusion does not occur until 1800 °C for SiC [48]. In addition to these properties, SiC is also very chemically stable making it resistant to erosion and corrosion. Altogether, these properties make SiC a prime candidate for use in harsh environments.

SiC is inert to most chemicals at room temperature though some reactions can happen at very high temperatures [49]. For example, SiC does not react with potassium hydroxide (KOH) at room temperature but will readily react around 600 °C with molten KOH. SiC is resilient to most chemical etchants used in the standard microfabrication environment; however, amorphous SiC can be etched with 1:1 HF:HNO3. Because only amorphous SiC reacts with this acid mixture, not single crystalline SiC, the HF:HNO_3 mixture is used for selective etching of amorphous SiC over single crystalline SiC. A recent report shows that this acid mixture can slowly etch poly-SiC as well; the etch rate can vary from 0.01 to 5nm/min. The etch rate is inversely proportional to grain size [50].

For most application environments described in this chapter, a high oxidative environment is common. SiC is highly resilient to atmospheric conditions and forms a negligible amount of oxide on the surface. SiC undergoes oxidation with oxygen, carbon dioxide, and steam at high temperature. The oxidation reactions are shown in equation (1.1), (1.2) and (1.3) [69]:

$$SiC(s) + 3/2O_2(g) \rightarrow SiO_2(s) + CO_2(g) \tag{1.1}$$

$$SiC(s) + 3H_2O(g) \rightarrow SiO_2(s) + CO(g) + 3H_2(g) \tag{1.2}$$

$$SiC(s) + 3CO_2(g) \rightarrow SiO_2(s) + 4CO(g) \tag{1.3}$$

Many qualitative and quantitative studies have been done to compare the oxidation characteristics of SiC under atmospheric conditions and in hydrocarbon combustion environments. It is found that SiC forms a silicon dioxide passivation layer and the rate of oxidation formation is diffusion limited at temperatures below 1200 K. For example, heating of poly 3C-SiC at 1025 K in atmospheric air resulted in a 5 nm thick oxide layer during the first five hours [69]. Going to even higher temperatures does eventually lead to an oxidation condition that results in significant material loss in SiC. At temperatures above 1200 K (927 °C), which is beyond the harsh environment application space discussed previously, competing redox reactions can occur in the presence of hydrogen, CO, and vapor H_2O forming volatile SiO [52]:

$$SiO_2(s) + H_2(g) \rightarrow SiO(s) + H_2O(g) \tag{1.4}$$

$$SiO_2(s) + CO(g) \rightarrow SiO(s) + CO_2(g) \tag{1.5}$$

$$SiO_2(s) + H_2O(g) \rightarrow SiO(OH)_2(g) \tag{1.6}$$

$$SiO_2(s) + H_2O(g) \rightarrow SiO(OH)_4(g) \tag{1.7}$$

This reduces the surface SiO_2 layer thickness resulting in a linear oxidation rate.

When comparing the oxidation behavior of common MEMS materials such as single crystalline silicon, diamond-like carbon (DLC) film, and poly 3C-SiC film under oxidation conditions relevant to the harsh environment applications of interest (1025 K in atmospheric air), SiC films possess a relatively high oxidation resistance. Under the above conditions, 500nm DLC was burned out during the first 24 hours

Table 1.2 Electrical properties of Si, GaAs, 6H-SiC, 4H-SiC, and 3C-SiC [53, 68]. When values depend on orientation to the c-axis (parallel, ||, or perpendicular, ⊥), both are listed.

Property	Si	GaAs	6H-SiC	4H-SiC	3C-SiC
Energy Bandgap [eV]	1.12	1.43	3.03	3.26	2.3
Thermal Conductivity [W/cm-K]	1.5	0.5	3.0-3.8	3.0-3.8	3-4
Intrinsic Carrier Concentration [cm^{-3}]	10^{10}	1.8×10^{6}	10^{-5}	10^{-7}	10
Saturated Electron Drift Velocity [10^7 cm/s]	1.0	1.2	2.0	2.0	2.5
Breakdown Field, Doping Conc. of 10^{17} cm^{-3} [MV/cm]	0.3	0.4	$\parallel$ 3.0 $\perp$ 2.5	$\parallel$ 3.2 $\perp$ 1.0	1.8
Electron Mobility	1200	6500	$\parallel$ 60 $\perp$ 400	800	750
Relative Dielectric Constant	11.9	13.1	9.7	10	9.6

while silicon and poly-SiC makes 290 and 48 nm of oxide after 100 hours, respectively [69]. This clearly demonstrates the superiority SiC over these other MEMS materials.

1.2.3 Electrical

In this section, two aspects of electrical properties are discussed. First, semiconductor material properties are discussed in terms of electronics that can survive high temperature, high power handling, and high radiation fields. Second, electrical properties are discussed in terms of a transducer material for high temperature environments. Mostly 4H-SiC and 6H-SiC are discussed in regards to electronics, as they are the polytypes currently available in wafer form. Single-crystal SiC and poly-SiC will be discussed as a piezoresistive sensing materials. Table 1.2 summarizes the basic electronics properties of SiC in addition to other commonly used semiconducting materials for comparison.

1.2.3.1 High temperature electrical behavior of SiC

High temperature electronics operation requires thermal stability of electrical parameters of the semiconductor itself as well as the semiconductor device. Thus, both intrinsic properties of the semiconductor and the device architecture play a critical role in temperature stability. This section focuses only on the intrinsic material properties aspect.

In semiconducting materials, increased temperature leads to a decrease in the energy band gap and an increase in carrier concentration, which in turn adversely affects the device performance. Materials with high bandgap and low intrinsic carrier concentration are needed for high temperature electronics applications.

As shown in Table 1.2, SiC possesses a higher energy bandgap compared to the more widely used Si and GaAs. Figure 1.11 graphically represents the estimated energy bandgap of Si, GaAs, 4H-SiC, and 6H-SiC as a function of temperature. The

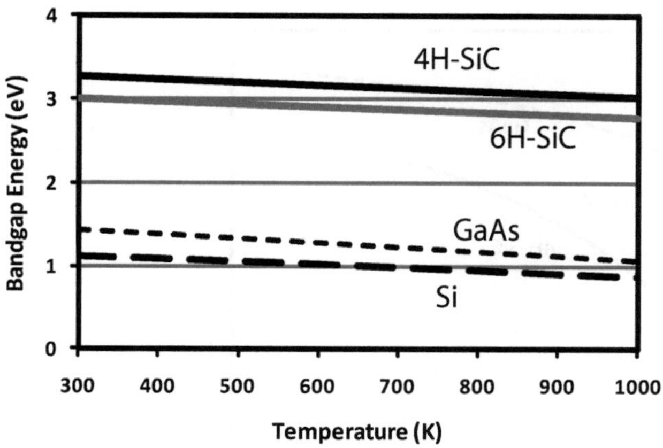

Fig. 1.11 Calculated energy band gap of Si, GaAs, and SiC as a function of temperature [54].

data clearly show a decrease in bandgap with increasing temperature. This bandgap reduction leads to larger intrinsic carrier densities, high leakage currents, poor junction rectification, and poor device isolation [54]. In comparison to Si and GaAs, both 6H-SiC and 4H-SiC retain a high bandgap value at even 1000 K, confirming SiC is better suited for high temperature operation.

Intrinsic carriers of semiconductors refers to thermal electrons and holes carriers present in the material at a given temperature. For proper operation of a semiconductor device, the intrinsic carrier density should be well below the intentional dopant-induced carrier density. For instance, the intrinsic carrier density of Si is 10^{10} cm^{-3} and well below the typical dopant-induced carrier densities for Si of 10^{14} to 10^{17} cm^{-3} at room temperature. For 6H-SiC, the intrinsic carrier density is 10^{-6} cm^{-3} and the typical dopant induced carrier density ranges from 10^{15} to 10^{17} cm^{-3} at room temperature; however, carrier concentration is strongly dependent on bandgap and temperature. The correlation of intrinsic concentration (n_i) and temperature is given by [54]:

$$n_i = [N_C N_V]^{1/2} exp[\frac{-E_g(T)}{2kT}] \tag{1.8}$$

where N_C and N_V are the number of carriers and vacancies respectively, $E_g(T)$ is the energy bandgap at a given temperature, and k is the Boltzmann constant.

As shown in Figure 1.12, the intrinsic carrier concentration of SiC stays well below allowable limits even at 1000K, allowing operation of SiC electronic devices at extremely high temperatures without suffering from intrinsic conduction effects.

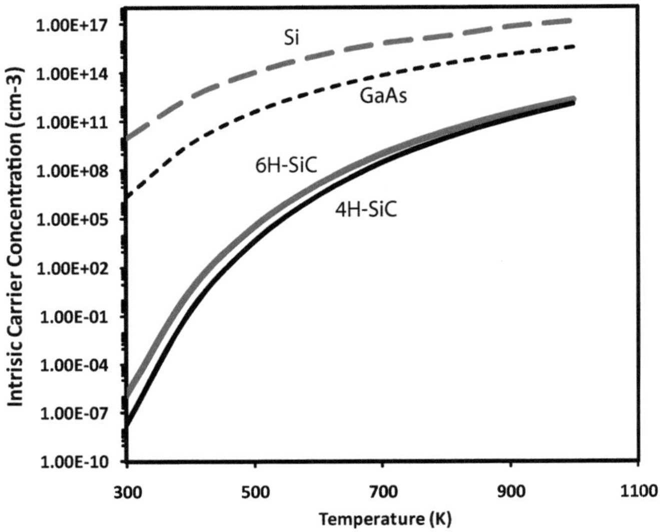

Fig. 1.12 Calculated intrinsic carrier concentration of Si, GaAs, and SiC as a function of temperature.

1.2.3.2 Wide-bandgap SiC for high power electronics

Solid-state semiconductor devices are being used or considered for power distribution and conversion. To fully tap the capabilities of solid-state devices, both materials and component level issues must be addressed. In this section, materials capability of SiC for power handling devices with comparison to common semiconducting materials will be discussed. More efficient high power electronics require low on resistance, low switching loss, high blocking characteristics, increased operating frequency, and high operational temperature capability. The primary material properties that improve these characteristics of high power solid state devices are large bandgap, high breakdown field, high saturated electron drift velocity, and high thermal conductivity.

SiC has a high breakdown field in comparison to silicon and gallium arsenide (Table 1.1), thus, SiC based power devices have higher breakdown voltages. A higher electric breakdown field also allows fabrication of device with thinner drift region.

The breakdown voltage (V_B) and the drift region width at a given breakdown voltage of a p-n diode can be approximated by equations 1.9 and 1.10, respectively [55]:

$$V_B \sim \varepsilon E_c^2/2qN_d \tag{1.9}$$

$$W(V_B) \sim 2V_B/E_c \tag{1.10}$$

Table 1.3 Comparison of breakdown voltage and drift region width of Si, GaAs, 4H-SiC and 6H-SiC.

Semiconductor	Si	GaAs	4H-SiC	6H-SiC
Breakdown voltage (V) for doping concentration N_d	$2.96 \times 10^{17}/N_d$	$5.76 \times 10^{17}/N_d$	$249 \times 10^{17}/N_d$	$274 \times 10^{17}/N_d$
Drift region width (cm) for breakdown voltage V_B	$6.67 \times 10^{-6}\ V_B$	$5.00 \times 10^{-6}\ V_B$	$0.67 \times 10^{-6}\ V_B$	$0.63 \times 10^{-6}\ V_B$

where ε is the permittivity ($\varepsilon = \varepsilon_r \varepsilon_0$), where ε_r is the dielectric constant and ε_0 is the vacuum permittivity) E_c is the breakdown voltage, q is the charge of an electron, N_d is the doping density. Using data from Table 1.2 and assuming the same doping density and drift region width, the breakdown voltage (V_B) of Si, GaAs, 4H-SiC, and 6H-SiC are shown in Table 1.3.

The data clearly shows that both 4H- and 6H-SiC devices have much higher breakdown voltages for a given doping concentration, and their drift regions are much thinner at a given breakdown voltage than their counterparts made from silicon and gallium arsenide. Thinner drift region yields lower resistance in the on state, resulting in much lower conduction losses. Moreover, SiC can be highly doped owing to its high breakdown field and can be utilized to further reduce the on state resistance. Reduction of on state resistance allows high power handling capability with better efficiency. As the device gets thinner, the storage of the minority carrier decreases, reducing the reverse recovery loss as well. That in turn enables high frequency operation. Furthermore, the switching frequency of SiC is much higher due to its high electron drift velocity.

High power handling and high switching frequencies of power devices increase the junction temperature. In comparison to Si and GaAs, SiC can operate in high temperature because of its wide bandgap as discussed earlier. Also, SiC has a much higher thermal conductivity in comparison to silicon and gallium arsenide that allows quick dissipation of generated heat to the environment, further lowering the temperature effect on the device.

1.2.3.3 Influence of radiation on SiC electrical properties

Semiconductors form acceptor or donor defects, also known as deep centers, upon exposure to radiation. These radiation-induced defects change the carrier density in the conduction band, which changes the conduction property of the material. For instance, in the case of acceptor defects, electron transfer from the conduction band to this defect results in a decrease in conductivity. Under extreme radiation doses, this process continues until a semiconductor eventually converts into an insulator. The process is commonly known as carrier (donor or acceptor) removal rate. The carrier removal rate of a semiconductor strongly depends on the threshold energy for defect formation under radiation. The threshold energy of a semiconductor can

Table 1.4 Calculated threshold energies of defect formation for Si, GaAS, Diamond, 3C-SiC, 4H-SiC and 6H-SiC.

Semiconductor	Si	GaAs	Diamond	3C-SiC	4H- and 6H-SiC
Lattice Constant, a_0 [Å]	5.65	5.431	3.57	4.36	3.08
Threshold Energy [eV]	9	12.8	80	37	153

be approximated using equation 1.11 [56]:

$$1.117E_d = (10/a_0)^{4.363} \tag{1.11}$$

where E_d is threshold energy of defect formation and a_0 is the lattice constant. The calculated threshold energies of defect formation for some common semiconductors are presented in Table 1.4.

Clearly from Table 1.4, 4H-SiC and 6H-SiC possess the highest radiation hardness among standard semiconductor substrate materials besides diamond, another contender for harsh environment electronics. Temperature also plays a vital role in radiation hardness of SiC as defect density decreases with increased temperature due to the high rate of recombination of primary defects at elevated temperatures.

1.2.3.4 Piezoresistance of SiC

Change in electrical resistance of a material when subjected to applied external stress is known as the piezoresistive effect. The effect has been used as a transduction mechanism for strain, pressure, and acceleration sensors. Typically the change in resistance is monitored by running a controlled current through the piezoresistor and monitoring voltage across the resistor. The sensor resistance changes due to an applied load, which causes a shift in voltage. The sensitivity of a piezoresistive sensor is quantified as gauge factor (GF), the percentage change in resistance per unit strain [57]. The relationship is given by equation 1.12:

$$GF = \frac{\Delta R}{R_0}\frac{1}{\varepsilon} \tag{1.12}$$

where R_0 is the nominal resistance of the network when no load is applied, ΔR is the change in resistance, and ε is the applied strain.

In addition, doping concentration, microstructure (single crystalline or polycrystalline), and operating temperature all have an effect on GF. At high temperature operation, piezoresistive transduction is affected by thermal stability, piezoresisitve gauge factor variation with temperature, and thermal coefficient of resistance of the material. Silicon is used in many piezoresistive transduction sensor applications owing to its high gauge factor [57, 58, 59]; however, poor thermal stability of silicon prevents its application to high temperature environments. SiC, in contrast, maintains a high gauge factor and is electrically and mechanically stable at high temperature.

Table 1.5 [60] Experimental room temperature gauge factor of n-type 3C-SiC at various resistivity levels and n-type silicon (©IEEE 1993), reprinted with permission.

Material	Resistivity	GF - π_{11}^*	GF - π_{12}^*	GF - $1/2(\pi_{11}+\pi_{12}+\pi_{44})^*$
n-type 3C-SiC	0.7	-31.8	19.2	-3.7
	0.02	-26.6		
	0.002	-12.7		
n-type silicon	11.0	133.0	68.3	-52

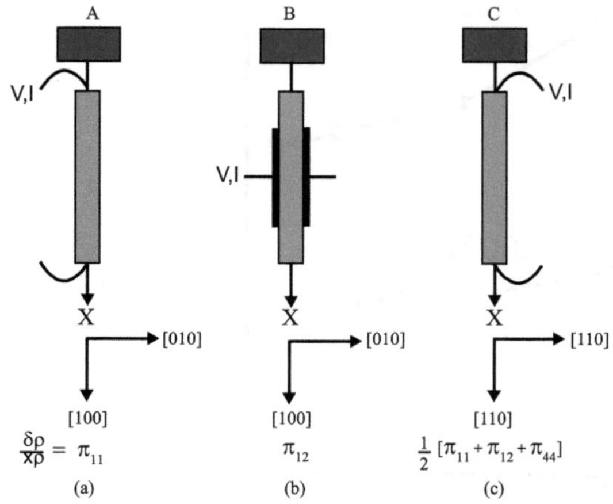

Fig. 1.13 [60] Schematic representation of strain gauges used to measure the GF of n-type 3C-SiC of different crystallographic orientations (©IEEE 1993), reprinted with permission.

Like silicon, the piezoresistance coefficient, and hence gauge factor, of SiC depends on the crystallographic orientation of the material. Table 1.5 lists the measured gauge factor of n-type 3C-SiC gauge factor at room temperature in comparison to n-type silicon. The GF corresponding to π_{11} has a larger value than the other coefficients, which means gauge orientation is a critical factor for maximizing sensitivity of 3C-SiC. On the other hand, the piezoresistance of 6H-SiC is isotropic within the base plane. Hence, the rotation of the gauge normal to the c-axis will not affect the gauge factor. Figure 1.13 shows schematically how metal foil strain gauges are placed to characterize the GF of different crystallographic orientations using a simple tension load test [60].

Although high gauge factor is important to improve strain sensitivity, thermal stability is also important to minimize aliasing of the sensor output to variations in temperature. For instance, Figure 1.14 shows the variation in GF for 3C-SiC with different resistivity values [60]. Although the higher resistivity layer exhibits a higher gauge factor, the lower resistivity layer maintains a decent GF yet exhibits very little shift in GF at operating temperatures above 500 K. This minimizes or eliminates the need for temperature compensation depending on the particular application requirements. Hence, 3C-SiC piezoresistive sensing is a viable load measurement technique in this temperature regime.

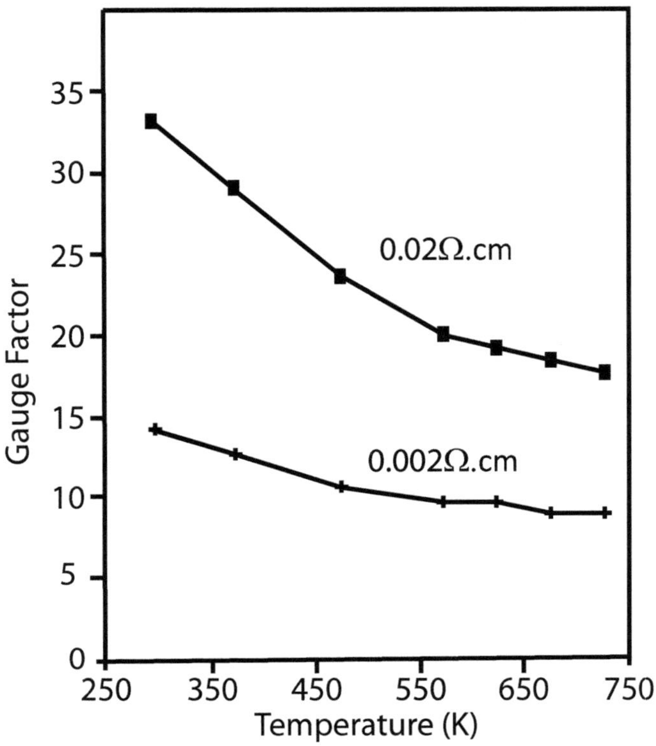

Fig. 1.14 [60] The gauge factor for single crystalline 3C-SiC with resistivity 0.02 and 0.002 Ω-cm as a function of temperature (©IEEE 1993), reprinted with permission.

1.2.4 Thermo-mechanical

Understanding mechanical properties is critical for designing sensors, actuators, and mechanical timing devices for applications in high temperature, high shock, and intense vibration conditions. Thermo-mechanical changes and shock- and vibration-induced stiction and fracture are considered the major failure mechanisms of micro-devices in the aforementioned conditions [61, 62, 63]. Thus, the thermo-mechanical stability, and the shock and vibration survivability are vital parameters in materials selection for harsh environment MEMS. This section reviews the thermo-mechanical properties of SiC.

The change in Young's modulus with temperature directly affects the performance of MEMS devices. This is true for static devices such as capacitive pressure and acceleration sensors as well as dynamic devices such as resonators. Based on recent studies of single crystalline 3C-SiC, the temperature coefficient of Young's modulus (TCYM) between room temperature and 800K is given by [66]:

$$E(T) = (-1.3 * 10^{-8}T^2 - 4.1 * 10^{-5}T + 1.0134)E_0 \tag{1.13}$$

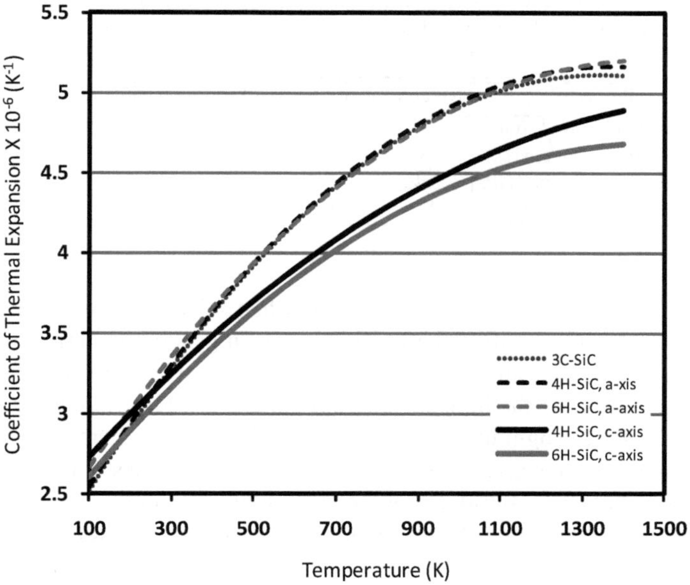

Fig. 1.15 CTE as a function of temperature of 3C-, 4H-, and 6H-SiC along their principle axis.

where E_0 is the Youngs modulus at room temperature.

Plastic deformation and changes in Young's modulus with temperature are the major parameters to consider when selecting materials for high temperature. The widely used MEMS material, Si, undergoes plastic deformation around 500 °C under minimal mechanical loads [64]. Hence, if high temperature operation is a requirement, silicon is not a viable option. Experimental data for 6H-SiC shows that macroscopic dislocation motion and plastic deformation starts above 1200 °C [65].

Furthermore, it is important to consider thermal expansion of SiC with changes in temperature (captured as a coefficient of thermal expansion, CTE). It is important because this expansion with temperature alters geometries or leads to stress changes in built-in beams when multiple material layers are used to form the device structure. In turn, this tends to change the response characteristics of a microstructure, aliasing the sensor response to the desired measurand. It is also important to note that often table values list a single value when typically CTE is a function of temperature. Figure 1.15 plots the CTE of 3C-, 4H-, and 6H-SiC along their principle crystal axes as a function of temperature [75]. Note that the CTE of SiC is significantly different whether the application is for room temperature or 600 °C, a shift in CTE of over 50%. It is also interesting to note that near room temperature the CTE of the various polytypes of SiC are actually close to the value for silicon.

Shock and vibration survivability is a fundamental requirement for most military and aerospace applications. MEMS devices can fail during shock and vibration due to two factors, stiction and fracture. Key material properties related to shock survival

Table 1.6 Bulk mechanical properties of select MEMS materials at 300 K [2, 63]. Note that some of the property values are from bulk material data. However, for feature size scales greater than 1 μm, the bulk material properties will be reasonably representative of the thin-film.

Property	Silicon	Silicon Nitride	Diamond	SiC
Young's modulus, E [GPa]	190	304	1035	448
Density, ρ [kg/cm^3]	2330	3300	3510	3300
Fracture Strength, σ_F [GPa]	2-4	5-8	8-10	4-10
E/ρ [GN/kg-m]	72	92	295	130

are listed in Table 1.6. Shock-induced stiction occurs when adjacent microstructures come into mechanical contact during the loading event. In general, materials with high E/ρ (stiffness-to-weight ratio) have reduced probability of stiction or fracture during a shock event because the high stiffness minimizes the defection while the low density decreases the magnitude of the inertial load. In terms of fracture failure, higher fracture strength is pivotal. SiC possesses higher fracture strength when compared to silicon and silicon nitride, only second to diamond; however, diamond is not as well suited for high temperature oxidative environments, as discuss previously.

1.3 System integration aspects of SiC materials

As discussed earlier in this chapter, SiC is an electronic semiconductor with a wide bandgap and high thermal conductivity, making it suitable for high power and high temperature operation. The wide band gap also reduces its sensitivity to radiation-induced damage. As a mechanical material for harsh environment MEMS, SiC possesses outstanding material properties, including high elastic stiffness and fracture toughness over silicon as well as stability of these properties beyond 500 °C. It also is chemically inert and resistant to wear. As will be discussed throughout the remainder of the book, SiC can be produced in various forms. They generally have similar mechanical and chemical properties yet have a variety of electrical properties and deposition temperatures, which will enable fabricating a variety of circuit elements and sensor structures.

Depending on the particular application and corresponding environmental constraints, integration complexity varies widely. Integration can be as simple as a micromachined device with an impedance matching buffer. It can be as complex as a micromachined device with control, sense, data processing, and communications electronics as well as on-board power supply. Integration of all major components of a microsystem (*i.e.*, electronics, sensors, and actuators) improves performance by reducing parasitics, increasing measurement sensitivity, reducing overall system size, and generally reduces power consumption. This high level of integration is also beneficial in reducing interconnect components, which further increases the opera-

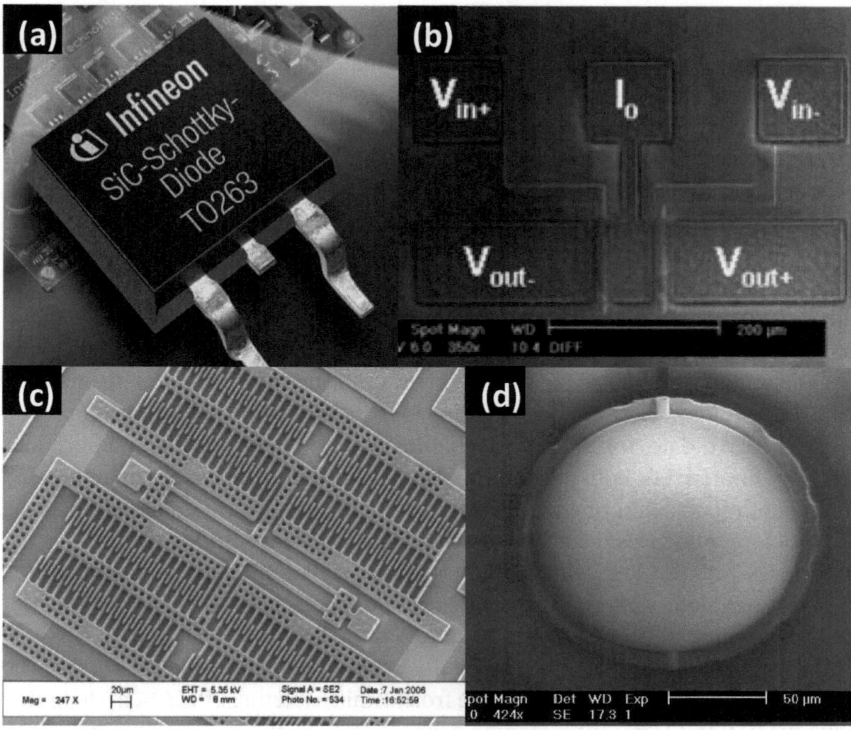

Fig. 1.16 (a) [70] picture of a commercial SiC Schottky diode from Infineon Inc., (b) [71] SEM micrograph of a 6H-SiC JFET based differential pair, (c) surface-micromachined 3C poly-SiC strain sensor, and (d) [67] SEM micrograph of a PECVD SiC encapsulated pressure sensor (©Elsevier 2003, Wiley-VCH Verlag GmbH & Co. KGaA 2009, and IOP 2004, respectively), reprinted with permission.

tional survivability in harsh environments since traditional interconnect technology is not typically suited for operation in corrosive or high temperature environments.

Figure 1.16 shows examples of discrete SiC devices fabricated using different forms of SiC (single-crystalline, epitaxial, polycrystalline and amorphous). These examples represent all major components of a microsystem: electronics, sensors, actuators, and packaging. From the materials point of view, not only the component but also the isolation and electrical routing layers must be fabricated using compatible materials. Compatibility in this case takes on several forms. First, the deposition methods must not damage previous layers, typically by not requiring excessively high deposition temperatures. Second, the coefficient of thermal expansion (CTE) differences between the various layers must be small because of the large operating temperature range expected for many of these harsh environment applications. Finally, both electrically insulative and conductive layers are needed and must meet the first two requirements in addition to withstanding the corrosive nature of the application in question.

Whether for high temperature operation or operation with frequent thermal cycling, one of the key areas of concern is CTE matching of every layer both at the discrete device level and system level. Thermal expansion differences between each layer as well as the substrate result in stresses during both fabrication and operation. In extreme cases, high thermal stress can lead to failure of the system due to delamination and cracking of the device layers [72]. These thermal stresses can also result in signal drift (unwanted thermal sensitivity). When the temperature-induced effects are large, passive or even active thermal compensation methods may not be able to successfully mitigate the problem [73, 74]. If the entire microsystem — substrate, electronics (including dielectrics), sensor components, and encapsulation — is produced only in SiC that would substantially reduce thermal mismatch as most forms of SiC have fairly similar CTE values. Thus, from the point of view of integrating a complete microsystem understanding CTE behavior of the complete material set is needed. To date, information available on CTE of poly-SiC and amorphous SiC is very limited. For both poly-SiC and amorphous SiC, deposition methods and specific conditions determine the microstructure of the resulting material, and the microstructure influences the CTE behavior. Thus, it is imperative to establish CTE data for poly-SiC and amorphous SiC based on deposition method and condition. This is one particular materials aspect that future research should focus on in order to realize complete SiC microsystem technology.

SiC can be produced in both electrically conductive and insulative forms. The conductivity of a SiC substrate can range from semi-insulating (10^5 Ω-cm) to highly conductive (0.028 Ω-cm) [68]. Doping concentration of the epitaxial SiC can be as high as $10^{19}/cm^3$. Poly-SiC can be produced with resistivity down to 0.02 Ω-cm while highly insulative amorphous SiC can be obtained by PVD and CVD methods [69]. Thus, it can be used as the primary platform for both mechanical device and electrical routing layers.

Utilizing the various forms of SiC for all aspects of the system would provide thermal, chemical, and mechanical stability for the entire system. Other materials with closely matched properties can also be considered for electrical and isolation layers. For example, silicon nitride and aluminum nitride are other possible isolation layer options for harsh environment applications since they both have closely matched CTE to SiC, can withstand high operating temperatures, and are fairly resistant to corrosion, although less so for silicon nitride.

This outlines the semiconductor technology aspects to integration. Integration of the many components goes well beyond the semiconductor properties alone. Issues such as compatibility of deposition and etching processes and thermal cycle influences on the electrical behavior of previously fabricated layers will be discussed along with fabrication techniques for SiC. Corrosion resistant metals that can make ohmic contact with SiC will be reviewed along with SiC-based electronics. Component interconnects will be explored when discussing SiC as a packaging material. At the end of the book, an overall approach is proposed that will take these various aspects of integration together as a whole to propose different integration schemes including a highly monolithic integration approach that simplifies piecewise integration of components and reduces system vulnerabilities to corrosive environments.

References

1. Kroetz GH, Eickhoff MH, Moeller H (1999). Silicon Compatible Materials for Harsh Environment Sensors. Sensors and Actuators 74:182–189
2. Mehregany M, Zorman CA, Rajan N, Wu CH (1998). Silicon Carbide MEMS for Harsh Environments. Proceedings of the IEEE 86(8):1594–1610
3. Hunter GW, Neudeck PG, Okojie RS, Beheim GM, Powell JA, Chen L (2003). An Overview of High-Temperature Electronics and Sensor Development at NASA Glenn Research Center. Journal of Turbomachinery 125:658–664
4. Sarro PM (2000). Silicon Carbide as a New MEMS Technology. Sensors and Actuators 82:210–218
5. Wright NG, Horsfall AB (2007). SiC Sensors: A Review. Journal of Physics D: Applied Physics 40:6345–6354
6. Hamada K (2009). Present Status and Future Prospects for Electronics in EVs/HEVs and Expectations for Wide Bandgap Semiconductor Devices. Materials Science Forum 600-603:889–893
7. Hillion M, Chauvin J, Grondin O, Petit N (2008). Active Combustion Control of Diesel HCCI Engine: Combustion Timing. USA Society of Automotive Engineers, Inc., Warrendale, PA:Report number 2008-01-0984
8. Yoon M, Lee K, Sunwoo M (2007). A Method for Combustion Phasing Control Using Cylinder Pressure Measurement in a CRDI Diesel Engine. Mechatronics 17:469–479
9. Toyota Motor Sales, USA Inc. Emissions #1 – Combustion Chemistry. http://www.autoshop101.com/forms/h55.pdf
10. Jurger RK (1999). Automotive Electronics Hadbook, McGraw-Hill, USA
11. Turner J (2009). Automotive Sensors. Momentum Press, New York
12. DeLaat JC, Chang CT (2003). Active Control of High Frequency Combustion Instability in Aircraft Gas-Turbine Engines. 16th International Symposium on Airbreathing Engines. Cleveland, Ohio, August 31-September 5, 2003:SABE2003–1054
13. He B, Shen T, Junichi K, Minggao O (2008). Input Observer-Based Individual Cylinder Air-Fuel Ratio Control: Modelling, Design and Validation. IEEE Transactions on Controls Systems Technology 16 (5):1057–1065
14. Minor RR, Rowe DW (1998).Utilization of GPS/MEMS-IMU for Measurement of Dynamics for Range Testing of Missiles and Rockets. Position Location and Navigation Symposium, IEEE 1998:602–607
15. Baum GA (1998). Manufacturing Process Control for Industry of the Future. National Academy Press, Washington, D.C. Publication NMAB-487-2
16. Kersey D (2000). Optical Fiber Sensors for Permanent Down Well Monitoring Applications in the Oil and Gas Industry. IEICE Transactions Electronics E83-C(3):400–404
17. Vandelli N (2008). SiC MEMS Pressure Sensors For Harsh Environment Applications. MicroNano News, April, 2008:10–12
18. Tschulena G (1988). Sensors for Process Control. Physica Scripta T23:293–298
19. Schadow KC (2004). MEMS Aerospace Applications. NATO Research and Technology Organization. RTO-EN-AVT-105
20. Brown TG, Davis B, Hepner D, Faust J, Myers C, Muller P, Harkins T, Hollis M, Miller C, Placzankis B (2001) Strap-Down Microelectromechanical (MEMS) Sensors for High-G Munition Applications. IEEE Transactions on Magnetics 37(1):336–342
21. Habibi S, Cooper SJ, Stauffer J-M, Dutoit B (2008). Gun Hard Inertial Measurement Unit Based on MEMS Capacitive Accelerometer and Rate Sensor. Position, Location and Navigation Symposium, 2008 IEEE/ION:232–237
22. Farrar CR, Worden K (2007). An Introduction to Structural Health Monitoring. Philosophical Transactions of Royal Society A 365:303–315
23. Eubank T (2007) Application of Condition Based Maintenance on Aerospace Structures. M.Sc. Thesis, Cranfield University
24. Romero R, Summers H, Cronkhite J (1996). NASA/CR-198446; ARL-CR-289

25. Beard SJ, Kumar A, Qing X, Chan HL, Zhang C, Ooi TK (2005) Practical Issues in Real-World Implementation of Structural Health Monitoring Systems. SPIE Smart Structures and Material Systems, San Diego CA, March 6-10, 2005:196–203
26. Cheng H (2007). Strategy for Assessment of WWER Steam Generator Tube Integrity. International Atomic Energy Agency. Report IAEA-TECDOC-1577
27. Kim I-S, Hong J-K, Kim H-N, Jang K-S (2003). Wear Behavior of Steam Generator Tubes in Nuclear Power Plant Operating Condition. Transactions of the 17th International Conference on Structural Mechanics in Reactor Technology (SMiRT 17), Prague, Czech Republic, August 1722, 2003:D04-5.
28. Wang GW, Pran K, Sagvolden G, Havsgard GB, Jensen AE, Johnson GA, Vohra ST (2001). Ship Hull Structure Monitoring Using Fibreoptic Sensors. Smart Materials and Structures 10:472–478
29. Chen H, Cardone V, Lacey P (1998). Use of Operation Support Information Technology to Increase Ship Safety and Efficiency. SNAME Transactions 106:105–127
30. Paik BG, Cho SR, Park B-J, Lee D, Yun J-H, Bae B-D (2007). Employment of Wireless Sensor Networks for Full-Scale Ship Application. IFIP International Federation for Information Processing, EUC 2007, LNCS 4808:113–122
31. Baldwin C, Kiddy J, Salter T, Chen P, Niemczuk J (2002). Fiber Optic Structural Health Monitoring System: Rough Sea Trials of the RV Triton. Oceans MTS/IEEE 3(3):1806–1813
32. Boller C (2001). Ways and Options for Aircraft Structural Health Management. Smart Materials and Structures 10:-432440
33. Gerardi TG (1990). Health Monitoring Aircraft. Journal of Intelligent Material Systems and Structures 1:375–384
34. Woelcken P, Bockenheimer C, Speckmann H, Entelmann W (2006). Outline of Overall Aircraft Imposed Requirements on Airframe Enhancements by Nanotechnologies and Resulting Opportunities. Proceedings of CANEUS 2006, August-September, Toulouse, France:69–72.
35. Staszewski WJ, Mahzan S, Traynor R (2009). Health Monitoring of Aerospace Composite Structures – Active and Passive Approach. Composites Science and Technology 69(11-12):1687–1685
36. Mancini S, Tumino G, Gaudenzi P (2006). Structural Health Monitoring for Future Space Vehicles. Journal of Intelligent Materials Systems and Structures 17:577–585
37. Derriso MM, Chang FK (2006). Future Roles of Structural Sensing for Aerospace Applications. NATO Research and Technology Organization: RTO-MP-AVT-141
38. Miller LM (1999). MEMS for Space Applications. SPIE Proceedings 3680:2–11
39. Tessler, A. (2007). Structural Analysis Methods for Structural Health Management of Future Aerospace Vehicles. NASA Report NASA/TM-2007-214871
40. Tor-Arne Grönland T-A, Pelle Rangsten P, Nese M, Lang M (2007). Miniaturization of Components and Systems for Space Using MEMS-Technology. Acta Astronautica 61:228–233
41. de Rooij RF, Gautsch S, Briand D, Marxer C, Mileti G, Noell W, Shea H, Staufer U, van der Schoot B (2009). MEMS for Space. Transducers 2009, Denver, CO, USA, June 21-25, 2009
42. Takahashi K (2004) Micro Thrusters for Miniaturized Space Systems, Need and Perspective. Power MEMS, Kyoto, Japan, Nov. 28-30, 2004:2–3
43. George T, Son KA, Powers RA, del Castillo LY, Okojie R (2005). Harsh Environment Microtechnologies for NASA and Terrestrial Applications. IEEE Sensors:1253-1258
44. Hunter GW, Okojie RS, Krasowski M, Beheim, GM, Fralick G, Wrbanek J, Greenberg, P, Neudeck PG, Xu J (2007). Microsystems, Space Qualified Electronics, and Mobile Sensor Platforms for Harsh Environment Applications and Planetary Exploration. 5th International Planetary Probe Workshop, Bordeaux, France, June 25-29, 2007.
45. Choyke WJ, Matsunami H, Pensl G (2004). Silicon Carbide: Recent Major Advances. Springer-Verlag, Berlin, Heidelberg, New York
46. Wijesundara MBJ, Valente G, Ashurst WR, Howe RT, Pisano AP, Carraro C, Maboudian R (2004). Single-Source Chemical Vapor Deposition of 3C-SiC Films in a LPCVD Reactor Part I: Growth, Structure, and Chemical Characterization. Journal of the Electrochemical Socciety 151:C210–C214.

47. Fu XA, Dunning JL, Zorman CA, Mehregany M (2005). Polycrystalline 3C-SiC Thin Films Deposited by Dual Precursor LPCVD for MEMS Applications. Sensors and Actuators A 119:169–176
48. Soloviev SI, Gao Y, Sudarshan TS (2000). Doping of 6H-SiC by Selective Diffusion of Boron. Applied Physics Letters 77(24):4004–4006
49. Zhuang D, Edgar JH (2005). Wet Etching of GaN, AlN, and SiC: a Review. Materials Science and Engineering 48:1–46.
50. Roper CS, Howe RT, Maboudian R (2009). Room-Temperature Wet Etching of Polycrystalline and Nanocrystalline Silicon Carbide Thin Films with HF and HNO_3. Journal of The Electrochemical Society 156(3):D104–D107
51. Wijesundara MBJ, Walther DC, Stoldt CR, Fu K, Gao D, Carraro C, Pisano AP, Maboudian R (2003). Low Temperature CVD SiC Coated Si Microcomponents for Reduced Scale Engines. Proceedings of ASME International Mechanical Engineering Congress and Exhibition, Washington D.C., November 15-21, 2003:IMECE2003-41696.
52. Fox DS, Opila EJ, Hann RE (2000). Paralinear Oxidation of CVD SiC in Simulated Fuel-Rich Combustion. Journal of American Ceramic Society 83(7):1761–1767
53. Neudeck PG (2006). Silicon Carbide Technology. The VLSI Handbook, Chapter 5 (Editor Wai-Kai Chen, CRC Press, Second Edition).
54. Patil AC (2009). Silicon Carbide JFET Integrated Circuit Technology for High-Temperature Sensors. Ph.D. Thesis. Case Western Reserve University.
55. Ozpineci B, Tolbert LM(2003). Comparison of Wide-Bandgap Semiconductors for Power Electronics Applications. ORNL/TM-2003/257
56. Lebedev AA, Kozlovski VV, Strokan NB, Davydov DV, Ivanov AM, Strel'chuk AM, Yakimova R (2002). Radiation Hardness of Wide-Gap Semiconductors (Using the Example of Silicon Carbide). Semiconductors 36(11):1270–1275
57. Kon S, Oldham K, Horowitz R (2007). Piezoresistive and Piezoelectric MEMS Strain Sensors for Vibration Detection. Proc. of SPIE Vol. 6529, Sensors and Smart Structures Technologies for Civil, Mechanical, and Aerospace Systems:65292V-1
58. French PJ, Evans AGR (1989). Piezoresistance in Polysilicon and its Applications. Solid-State Electronics 32(1):1–10
59. Suhling JC, Jaeger RC. Silicon Piezoresistive Stress Sensors and Their Application in Electronic Packaging. IEEE Sensors Journal 1(1):14–30
60. Shor JS, Goldstein D, Kurtz AD (1993). Characterization of n-Type β-SiC as a Piezoresistor. IEEE Transactions on Electron Devices 40(6):1093–1099
61. Srikar VT, Spearing SM (2003). Materials Selection in Micromechanical Design: An Application of the Ashby Approach. Journal of Microelectromechanical Systems 12(1):3–10
62. Srikar VT, Senturia SD (2002). The Reliability of Microelectromechanical Systems (MEMS) in Shock Environments. Journal of Microelectromechanical Systems 11(3):206–214
63. Spearing SM (2000). Materials Issues in Microelectromechanical Systems (MEMS). Acta Metallurgica 48:179–196.
64. Yonenaga I (2003). High-temperature Strength of Bulk Single Crystals of III-V Nitrides. Journal of Materials Science: Materials in Electronics 14:279–281
65. Yonenaga I (2001). Thermo-Mechanical Stability of Wide-Bandgap Semiconductors: High Temperature Hardness of SiC, AlN, GaN, ZnO and ZnSe. Physica B 308-310:1150–1152
66. Pozzi M, Hassan M, Harris AJ, Burdess JS, Jiang L, Lee KK, Cheung R, Phelps GJ, Wright NG, Zorman CA, Mehregany M (2007). Mechanical Properties of a 3C-SiC Film Between Room Temperature and 600 °C. Journal of Physics D: Applied Physics 40:3335–3342
67. Pakula LS, Yang H, Pham HTM, French PJ, Sarro PM (2004). Fabrication of a CMOS compatible pressure sensor for harsh environments. Journal of Micromechanics and Microengineering 14(11):1478–1483
68. Cree Semiconductor Product Specifications, www.cree.com
69. Wijesundara MBJ, Gao D, Carraro C, Howe RT, Maboudian R (2003). Nitrogen Doping of Polycrystalline 3C-SiC Films Grown Using 1,3-Disilabutane in a Conventional LPCVD Reactor. Journal of Crystal Growth 259:18–25

70. Telford M (2003). SiC's power cuts cost. III-Vs Review 16(4):44–47
71. Neudeck PG, Gaverick SL, Spry DJ, Chen L-Y, Beheim GM, Krasowsk MJ, Mehregany M (2009). Extreme temperature 6H-SiC JFET integrated circuit technology. Physica Solidi A 206(10):2329–2345
72. Savrun E (2002). Packaging Considerations for Very High Tepmerature Microsystems. Sensors 2002, June 12-14, 2002:1139–1143
73. Vig JR (2001). Temperature-Insensitive Dual-Mode Resonant Sensors – A Review. IEEE Sensors Journal 1:62–68
74. Melamud R, Kim B, Hopcroft MA, Chandorkar S, Agarwal M, Jha CM, Kenny TW (2007). Composite Flexural-Mode Resonator with Controllable Turnover Temperature. MEMS 2007, Kobe, Japan:199–202
75. Li Z,Bradt C (1986). Thermal Expansion of the Cubic (3C) Polytype of SiC. Journal of Materials Science 21:4366–4368

Chapter 2
SiC Materials and Processing Technology

This chapter contains a broad review of SiC materials and processing technology necessary to create SiC electronics, micromechanical transducers, and packaging. Details on deposition and etching methods are covered. The material properties of various forms of SiC (single crystalline, polycrystalline, and amorphous) along with their use for creating the various components of harsh environment microsystems will also be discussed. Current status and future research are highlighted with regards to both materials and processing technologies.

2.1 Material considerations for various applications

The creation of harsh environment microsystems using SiC is advantageous because all of the system components can be made from SiC. Semiconductor grade SiC is commercially available for electronic device fabrication. MEMS structures can be fabricated using single-crystalline, poly-crystalline, or amorphous SiC. Likewise, packaging of MEMS and electronics can be accomplished using any of these forms of SiC. The following sections will briefly describe the materials consideration specific to each of these components of the microsystem.

2.1.1 Electronics (crystallinity, doping, defects, polytype)

As described in Chapter 1, over 200 SiC polytypes exist. Among all the polytypes, 3C-, 4H-, and 6H-SiC are the most commonly available today. Each SiC polytype exhibits different electrical, optical, and thermal properties due to differences in stacking sequence. Some of the key electrical parameters for 3C-, 4H-, and 6H-SiC are listed in Table 2.1 (a more detailed list of properties of SiC polytypes can be found in Table 1.2 in Chapter 1). The significant electrical disparity among

Table 2.1 Key electrical parameters of SiC [1]

Property	4H-SiC	6H-SiC	3C-SiC
Bandgap [eV)]	3.2	3.0	2.3
Intrinsic Carrier Concentration (cm^{-3})	10^{-7}	10^{-5}	10
Electron Mobility at $N_D = 10^{16}$ (cm^2/V-s)	‖ c-axis: 800 ⊥ c-axis: 800	‖ c-axis: 60 ⊥ c-axis: 400	750
Hole Mobility at $N_D = 10^{16}$ (cm^2/V-s)	115	90	40
Donor (nitrogen) Dopant Ionization Energy (MeV)	45	85	40

these polytypes clearly shows that it is essential to use a single polytype (single-crystalline) for electronics device fabrication.

4H- and 6H-SiC are the only choice for wafer substrates since 3C-SiC wafers are not yet commercially available. Regardless of polytype, fabrication of devices directly on SiC wafers is hindered by lack of device quality wafers, inability to drive in surface doping (Chapter 1.1.2), and poor electrical quality as a result of direct ion implantation into the substrate [2]. Therefore, the SiC electronics fabrication is mainly centered on epitaxial layers grown on these substrates. Currently, high quality homoepitaxial layers of 4H- and 6H-SiC with different thicknesses and doping levels are routinely produced. 3C-SiC is also gaining attention as it can be grown heteroepitaxially on various substrate materials. Furthermore, there has been significant progress in producing device-grade 3C-SiC epilayers in recent years. However, it is necessary to further reduce crystallographic structural defects in 3C-SiC epilayers before this polytype becomes a viable alternative to 4H- and 6H-SiC [3].

Low defects, controlled doping, and dopant uniformity of both substrate and epilayers are crucial for device applications. In terms of the device type, low-resistivity substrates are vital for power devices as they reduce power losses due to parasitic substrate and contact resistances [1]. However, for devices and circuits operating at microwave frequencies; it is necessary to have semi-insulating substrates to achieve low dielectric losses and reduced device parasitics.

Most current SiC based electronics devices are fabricated using either 4H- or 6H-SiC due to the aforementioned short coming of 3C-SiC. Between 4H- and 6H-SiC, 4H-SiC has substantially higher carrier mobility, shallower dopant ionization energies, and low intrinsic carrier concentration (Table 2.1). Thus, it is the most favorable polytype for high-power, high-frequency, and high temperature device applications. In addition, 4H-SiC has an intrinsic advantage over 6H-SiC for vertical power device configurations because it does not exhibit electron mobility anisotropy while 6H-SiC does [1]. Therefore, many SiC device fabrication efforts have shifted towards 4H-SiC as it has become more readily available.

2.1.2 MEMS (Stress, strain gradient)

Silicon carbide MEMS are fabricated using both single-crystalline substrates and thins films. When MEMS structures are created using single-crystalline substrates, bulk micromachining (discussed in Chapter 4) is used. Fabrication of devices directly from the substrate material has the inherent advantage of having the same mechanical and electrical properties of original substrate and material optimization for MEMS can be avoided, particularly removing the need for stress and strain gradient optimization. However, from the structural device standpoint, bulk micromachining is limited in terms of device architectures that can be realistically fabricated. Thus, most SiC devices are fabricated using surface micromachining techniques.

In surface micromachining, the structural SiC layer is typically deposited on a different material layer, such as a sacrificial or isolation layer, and this interlayer involvement leads to deviations of both electrical and mechanical properties compared to the substrate. At the current maturity of the technology, it is not possible to achieve single-crystalline SiC with optimized electrical and mechanical properties for MEMS on widely-used sacrificial or isolation layers *i.e.* polysilicon, silicon dioxide, and silicon nitride. However, many MEMS devices do not require as stringent electrical characteristics as are needed for electronics. This allows the use of electronically inferior but mechanically sound alternatives. For instance, polycrystalline SiC (poly-SiC) can be produced with comparable elastic stiffness to single-crystalline SiC and can have sufficient electrical characteristics for MEMS applications. To date, most SiC MEMS devices are fabricated with poly-SiC. Amorphous SiC, though mostly electrically insulative, has also been used for MEMS components such as diaphragms as well as for packaging structures.

The key concerns with amorphous SiC and poly-SiC thin films are the residual stress and stress gradient of the deposited films. The residual stress in thin films can be attributed to many different sources. The films are deposited at temperatures significantly above ambient, and the difference in thermal expansion coefficients between the film and the underlying layers leads to stress generation upon cooling to room temperature. Crystal defects, impurity incorporation, and grain growth and orientation are also sources of film stress. When the mechanical microstructure is released the residual stresses in the structural layers relieve by deforming the structural layers. In extreme cases, the thin film may crack while attached to the substrate or delaminate from the underlying layers. Furthermore, a stress gradient exists when the residual strain in a film varies through its thickness. This is particularly troublesome for MEMS applications because it can cause significant curvature of free-standing microstructures even when the average residual stress of the film is near zero.

MEMS typically demands significantly thicker films than is needed for electronics fabrication. This further complicates the desire to produce low stress and low stress gradient films. Changes in stress levels through the film thickness occur because of changes in grain size over the course of the deposition. Therefore, micromechanical sensor fabrication requires processes that can precisely control mechanical characteristics of deposited thin films. This will be described in more detail in Chapter 4.

2.2 Substrate

Production of semiconductor grade SiC ingots is one of the most challenging tasks faced by the SiC semiconductor and microsystems industry. Even though the first reported growth of SiC crystals by the Acheson technique dates back to 1892 [4], growth of single-crystalline SiC substrates with both size and quality equivalent to most common semiconductors — silicon, gallium arsenide, and indium phosphide — is yet to be realized. Difficulty growing single-crystalline SiC is attributed to the physical-chemical nature of SiC. The SiC phase diagram reveals a peritectic decomposition at 2830 °C and pressure around 10^5 Pa. Theoretical calculations indicate that stoichiometric melting can be attained when temperatures and pressures exceeding 3200 °C and 10^5 atm, respectively [56]. Therefore, growing single-crystalline SiC boules from stoichiometric melt or from solution (similar to common semiconductor growth) is impractical. Furthermore, using these very high temperature conditions makes producing single-crystalline materials rather demanding because of the narrow differences in enthalpy of formation of the different polytypes of SiC [6]. These reasons led SiC semiconductor developers to seek alternative methods that include solution phase growth from non-stoichiometric solutions and vapor phase growth. Solution phase routes exploit the solubility of SiC and carbon in silicon melt to grow SiC from non-stoichiometric solutions while vapor phase growth utilizes the sublimation products of SiC or CVD methods to produce SiC substrates.

The most common SiC substrate process is based on the vapor phase growth using sublimation products of SiC. SiC sublimes at temperatures above 1800 °C, forming various elemental and molecular species. The sublimation not only produce SiC(gas) but also forms various elemental and molecular species due to dissociation of SiC as shown in the chemical equations below [7]:

$$SiC(s) \rightarrow Si(g) + C(s) \tag{2.1}$$

$$2SiC(s) \rightarrow SiC_2(g) + Si(g) \tag{2.2}$$

$$2SiC(s) \rightarrow C(s) + Si_2C(g) \tag{2.3}$$

The first successful growth of high purity SiC crystal using sublimation growth was developed by Lely in 1955 [8]. Lely's initial process used a dense graphite crucible and a porous graphite thin-walled inner cylinder. The SiC powder is loaded between the inner and outer cylinder and heated to temperatures of 2550-2600 °C in an argon atmosphere. Spontaneous nucleation of SiC was observed at the inner surface of the thin-walled cylinder. The Lely method had limited impact on SiC crystal growth due to two main reasons. First, the control of platelet thickness, doping, and polytype of the crystal is poor due to the inability of controlling initial nucleation, growth rate, and growth direction. Second, Lely method is capable of producing SiC crystals only up to 10 mm in diameter due to self-termination of the growth process as pores of the inner cylinder are blocked by growing crystals.

A major breakthrough came around 1980 with the introduction of SiC seed crystal sublimation growth. This method uses a high quality seed crystal surface to begin growth process in contrast to Lely's method which begin growth on a graphite membrane [9]. The seeded sublimation growth is often referred to as the physical vapor transport method (PVT) or Modified Lely Method. Later, this method was further refined for producing large diameter SiC boules [10, 11, 12]. Various improvements and modifications of PVT evolved from both commercial production environments and research laboratories until today it has become the current standard industrial process. High quality SiC wafers are routinely produced with the current PVT method (often referred to as the standard PVT method). Currently 100 mm 4H- and 6H-SiC wafers are commercially produced by standard PVT and 150 mm wafers are expected near future [14, 15].

The ever increasing demand for further improvement of quality and production throughput of SiC wafers requires either additional improvement of standard PVT or exploring other viable options. This has led to many successful attempts to fuse CVD based methods or further modify PVT techniques for producing semiconductor grade SiC wafers. One such method gaining attention for industrial wafer production applications is high temperature CVD (HT-CVD) [16].

Other recent efforts include continuous feed PVT (CF-PVT) [17], Halide CVD (H-CVD) [18], and Modified PVT (M-PVT) [19]. While these techniques claim to have some technological edge over their predecessors, they are still primarily at the research stage. Solution phase growth has yet to prove it is capable of producing large area substrates. Nonetheless, promising initial results and the advantages of this method will certainly draw more attention from the research and industrial community. Figure 2.1 summarizes the current status of known SiC substrate technology. The following section will briefly describe the processes involved in each of these techniques, and discuss their relative advantages over the well-established standard PVT method.

2.2.1 Vapor Phase Growth

2.2.1.1 Standard Physical Vapor Transport (PVT) or Seeded Sublimation Method

Standard PVT growth is carried out in a quasi-closed graphite crucible in an argon environment. The reactor configuration is schematically shown in Figure 2.2. The source material is held at the bottom of the crucible, and the seed is fixed onto the top lid of the crucible. The distance between source and seed is typically around 20 mm [20]. The crucible is inductively heated and operating temperatures of the system ranges from 1800 to 2600 °C. A linear temperature gradient is maintained

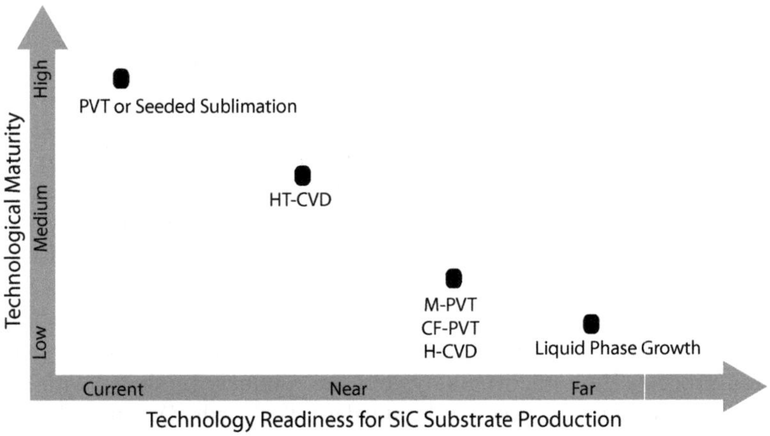

Fig. 2.1 The current state of possible SiC substrate technology available today.

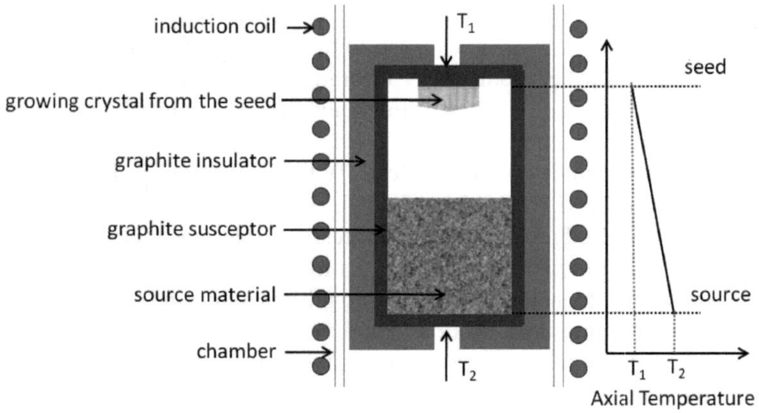

Fig. 2.2 The schematic representation of the standard PVT reactor with representative axial temperature profile.

between source and the seed causing vapor phase transport of Si and C containing species from the source to the seed. The typical temperature gradient ranges from 1.0 to 2.5 °C per mm. In most cases, the temperature at the bottom of the crystal is around 2300 °C while the seed temperature is around 2100 °C. The crystallization process is facilitated by the supersaturation of a vapor phase species at the seed surface [14, 15].

SiC boule growth via the sublimation process is complex and many growth parameters have to be controlled precisely for high quality crystal formation [14]. The suggested profile of two key parameters, temperature and pressure, during the growth process is shown in Figure 2.3. Initially, the source material is preheated up to 1800 °C. High argon pressure, typically around 600 Torr, is held during pre-

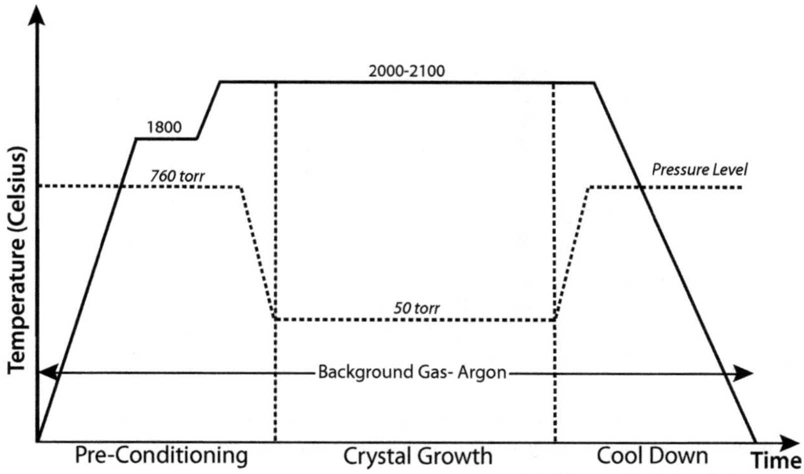

Fig. 2.3 [20] Temperature and pressure profile of the PVT growth process (©Taylor & Francis 1997), reprinted with permission.

heating and until the growth temperature stabilizes over 2200 °C to eliminate low temperature polytype growth, particularly 3C-SiC. Once at growth temperature, the argon pressure is reduced below 50 Torr to initiate the growth process [20]. Background argon pressure is again increased above 600 Torr to stop the growth process before decreasing the temperature. The typical growth rate of seeded sublimation lies between 0.2 to 2 mm per hour [14].

The nucleation, growth, and defect formation are intertwined with the temperature profile of the reactor and boule. Slight variations in the temperature profile can lead to defect formation via polytype formation and elastic deformation. Therefore, temperature control is a key aspect of this process. Defects in the seed crystal surface directly translate into defects in the bulk crystal. Hence, a high quality seed crystal is also critical for creating a high quality SiC substrate.

Other factors that are highly critical for SiC boule growth is the control of polytype and doping concentrations. Recently, several methods have been developed for controlling polytype and doping levels of SiC boules. Some of these technologies will be detailed in the later section of this chapter. They include the use of the surface polarity of the seed crystal and introduction of rare-earth elements to the source material [21, 22].

To date, the standard PVT method has become the dominant SiC growth technique used in the SiC industry. There is vast knowledge base in both academia and industry around standard PVT. This technique is capable of producing commercially-viable, large dimension SiC boules, up to 100 mm, with high crystal quality.

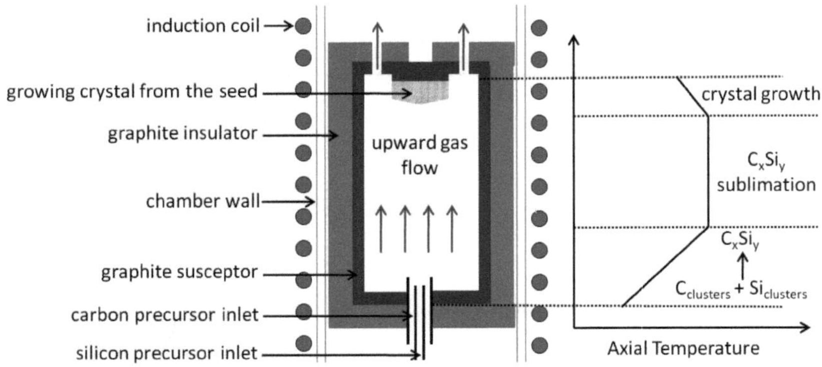

Fig. 2.4 [24] The schematic representation of the reactor geometry of HTCVD system with the temperature profile of the reactor along with corresponding chemical stages (©Trans Tech Publishing 2004), reprinted with permission.

2.2.1.2 High Temperature CVD Method (HTCVD)

The High Temperature CVD Method (HTCVD) for SiC bulk crystal growth is relatively young in comparison to PVT growth [23]. Nonetheless, tremendous progress has been reported in this method since its inception in 1996 [24]. HTCVD has very recently been used to produce commercial SiC wafers [15]. Figure 2.4 shows the schematic representation of the vertical reactor used in this process. The gas delivery components and retractable rotating crystal mount are the major differences between HTCVD and standard PVT reactors. Furthermore, the temperature profile of the reactor as shown in Figure 2.4 is also different from standard PVT.

The growth is carried out using conventional silicon and carbon containing precursors. Silane (SiH_4) is used as the silicon source while methane, ethane, or propane can be used as the carbon source. The growth precursors are introduced in coaxial tubes with the inner-most tube used for the silicon precursor. High flow rate and pressure conditions are maintained so that the dissociation of silane can lead to the formation of Si clusters by homogeneous gas phase nucleation. These clusters react with the hydrocarbon precursor as the temperature further increases downstream, forming Si_xC_y clusters. Upon entering to the high temperature heating zone, these Si_xC_y clusters sublimate to form Si and C containing species before reaching the seed crystal. Crystal growth occurs via supersaturation, similar to the PVT technique. The typical growth rate associate with this process is 0.1 to 0.7 mm/h.

The technique offers some intrinsic advantages over the standard PVT method. The continuous feed of the precursor allows direct control of the Si to C ratio as well as the dopant concentration. The availability of high purity gas ensures the purity of the crystals. Semi-insulating crystal growth using HTCVD has been demonstrated [25]. The versatility of this method has been shown by growing p-type substrates with precise dopant control [26]. Despite initial successes, further refinement is

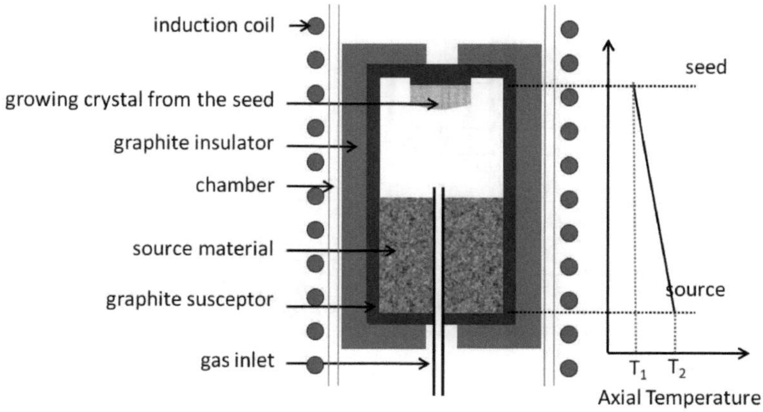

Fig. 2.5 [14] Typical reactor configuration of M-PVT method and the temperature profile of the reactor (©IOP 1997), reprinted with permission.

needed in order to obtain large diameter crystals, which would enable it to be a commercially-viable competitor to standard PVT.

2.2.1.3 Modified PVT Method

The modified PVT (M-PVT) method is derived from both the standard PVT and CVD methods. Figure 2.5 schematically illustrates the reactor architecture used in M-PVT along with the axial temperature profile of the reactor. The configuration is very similar to that of a conventional PVT reactor with the added capability of delivering small amounts of Si and C containing gaseous precursors, as well as dopant precursors to the system [19]. This configuration enables a growth process carried out in a very similar fashion to PVT but with more precise control of stoichiometry and dopant uniformity of the SiC crystals.

In a conventional PVT reactor, the gas phase composition of various Si and C containing gas species is determined by the temperature field that is set by the heating procedure and crucible design. However, the temperature field can change during the growth process due to process instabilities induced by evolution of the crystal as well as changes in gas phase composition arising from morphological changes in the source materials. In M-PVT, the ability to feed Si and C containing gases at the growth front allows minimizing the variations in gas phase composition, enabling better controlling growth of SiC crystals.

The possibility of continuous feeding of dopant gases also improves the resulting doping level uniformity in both axial and radial directions. In PVT, nitrogen is usually used for n-type doping. Nitrogen gas is incorporated into the wall of the growth crucible, exploiting the porosity of the graphite. In the case of p-type doping, aluminum is directly mixed into the SiC source material. The much higher vapor pressure of aluminum in comparison to Si and C containing gas species makes mass

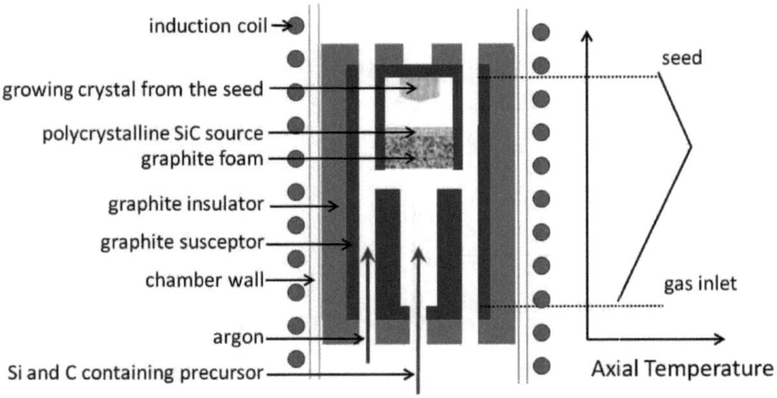

Fig. 2.6 [14] Schematic drawing of the reactor configuration of CF-CVD and the axial temperature profile (©IOP 1997), reprinted with permission.

transport control rather difficult. M-PVT enables continuous feed of aluminum vapor or aluminum containing precursor with fine control. This method may lead to the fabrication of low resistance p-type SiC wafers that will open paths to fabricate power SiC device on p-type wafers. Ion implantation experiments demonstrated that phosphorus exhibit a ten-fold increase in solubility limit compared to nitrogen [27]. Thus, it can be argued that higher n-type doping can be achieved using M-PVT by continuous feed of phosphine gas, a widely used n-type doping precursor in the semiconductor industry.

2.2.1.4 Continuous Feed PVT (CF-PVT)

The continuous feed PVT is essentially a hybrid of both PVT and HT-CVD. This method exploits the fundamental advantages of both PVT and HT-CVD techniques for producing high quality SiC crystals [17, 15]. Figure 2.6 shows the typical configuration of the CF-PVT reactor. A Si and C containing single precursor tetramethylsilane diluted in argon is typically used as the source material. At the low temperature zone, tetramethylsilane forms high quality poly-crystalline SiC through a process similar to HT-CVD. This poly-crystalline SiC source material is transferred to the high temperature sublimation zone through a highly porous graphite foam layer. The SiC growth occurs through supersaturation similar to that of classical PVT. The growth rate obtained by this method is around 100 μm/h at 1900 °C.

One of the inherent advantages to the CF-PVT is its ability to grow longer SiC ingots because of the continuous supply of the source material. In classical PVT, the supersaturation close to the seed surface is controlled by the pressure and temperature distributions within the crucible. This method adds another parameter to process control: precursor concentration. It has been shown that the feeding gas flow rate that controls the source material formation can be used for precise control

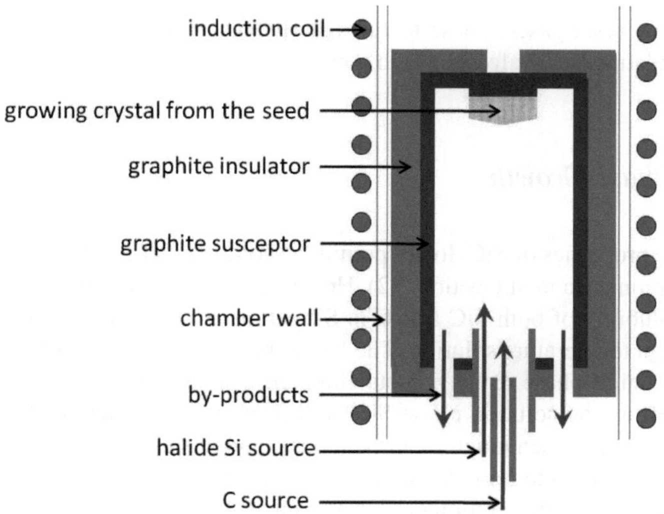

Fig. 2.7 [14] Schematic representation of the HCVD growth reactor (©IOP 1997), reprinted with permission.

of the supersaturation close to the seed that, in turn, enables a means of polytype control [28]. Although CF-PVT is at an early stage of development, it has successfully produced both high purity 4H-SiC and 3C-SiC [28].

2.2.1.5 Halide CVD (HCVD)

Halide Chemical Vapor Deposition (HCVD) is a recently introduced and very promising technique for the creation of high purity SiC crystals [18, 30]. The process is done in an inductively heated chamber very similar to a HT-CVD reactor; however, the growth mechanisms are fairly different. HCVD growth occurs through surface reaction, and the growth rate is determined by desorption kinetics. The process characteristics are the same as the conventional CVD. The earlier described HT-CVD growth takes place by a sublimation process, and growth rate is determined by supersaturation. Figure 2.7 schematically represents the reactor. A chlorinated Si precursor ($SiCl_4$), a C precursor (C_3H_8 or CH_4), and hydrogen feed upward from the bottom of the reactor via separate concentric graphite injectors. Chlorinated precursors are used to avoid homogeneous nucleation in the gas phase. Reactor temperature is maintained around 2000 °C, and no temperature gradient exists between seed and source in contrast to all other previously described method. The typical growth rate of this method is 250-300 μm/h.

HCVD possesses all the intrinsic advantages that classical CVD offers. The stoichiometry of SiC crystals can be easily controlled by using the flow rates of the precursors and keep constant throughout the growth process. The crystal grown with HCVD has very low impurity levels and high electrical resistivity mainly due to the use of high purity precursor gases. These features of HCVD are very attractive for producing semi-insulating SiC wafers for high power devices.

2.2.2 Solution Phase Growth

The thermodynamic properties of SiC do not permit the solution phase bulk growth of SiC from a stoichiometric melt (section 2.2). However, many studies have taken advantage of the solubility of both SiC and C in Si melt as an alternative route to grow SiC from a high temperature solution. The key to this method is that SiC can be grown from the liquid phase using non-stoichiometric solutions containing Si and C [31]. Early efforts on solution phase SiC crystal growth date back to 1961 [32]; however, there has not been much progress reported on this method until very recently. The main reason is the remarkable success of PVT methods, which have diverted attention away from the solution growth approach. The recent focus on liquid phase bulk growth of SiC is mainly driven by its ability to grow crystals with low dislocation densities and grow crystals at relatively low temperatures (1500-1700 °C at the seed). Solution growth process occurs under the conditions close to thermal equilibrium resulting in high quality crystals with better polytype controllability [33, 34]. The low growth temperature is also attractive for 3C-SiC growth as this polytype forms at relatively low temperatures, and sublimation growth, which requires relatively high growth temperatures, is not suitable for 3C-SiC growth.

The low solubility of C in Si melt, which is directly related to the growth rate, is one of the limiting factors to this method. This problem can be mitigated by adding transition metals into the Si melt, which increases the solubility of carbon for higher growth rate. Besides the solubility aspect, there are other important issues that must be considered when selecting the solvent (Si + metal) system. Those include no metal incorporation into the solid, excellent wetting of the crystal by the solvent, low vapor pressure, low melting point, and reduced degradation of the crucible due to reaction mixture. Furthermore, the crystal should be the only stable solid phase [31]. A few solvent systems have been successfully employed, and reasonable progress has been made in terms of increasing growth rate and the crystal diameter.

Several growth techniques have been considered for solution growth of SiC. The commonly discussed techniques include traveling zone method, slow cooling technique, and top seed solution growth method (TSSG) [31, 35, 36]. To date, TSSG is the most successful method of growing SiC crystals from solutions. Figure 2.8 shows a typical reactor configuration for TSSG growth. The seed crystal is mounted on a graphite rod inserted into the growth crucible. The seed and crucible can rotate with respect to each other. In typical process mode, only the seed is rotated at 10-20 rpm. In the accelerated crucible rotation technique, which is used for increasing growth rate, both the crucible and the seed are rotated in opposite directions. The maximum crucible and seed rotation rates are typically 20 and 10 rpm, respectively. The seed is held as the low temperature point and a temperature gradient of 2.0 °C/mm is maintained between the seed and the bottom of the crucible.

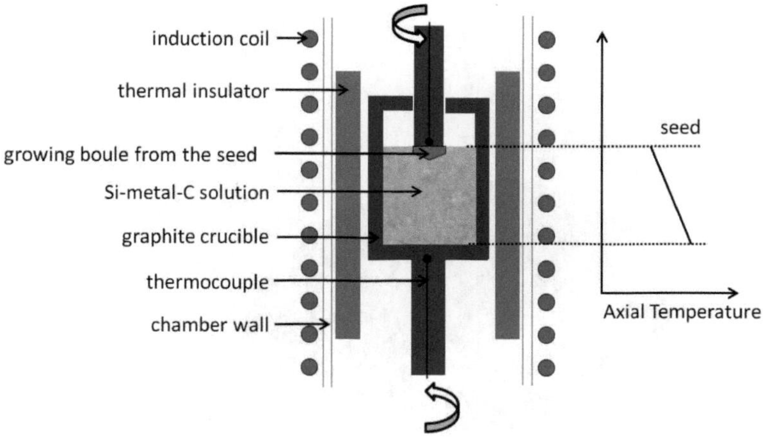

Fig. 2.8 [37] Schematic representation of the TSSG growth reactor (©Elsevier 2008), reprinted with permission.

Initial research done by Hoffmann *et al.* [31] using TSSG has demonstrated growth of 1.4 inch 6H-SiC crystals with a high degree of crystallinity. The reported growth rate ranges between 0.05-0.2 mm/h. Recent research reported from Japan shows the growth of 6H-SiC single crystals from Si-Ti-C ternary solution using TSSG. Two inch diameter SiC crystal with thickness of 5mm (Figure 2.9) was grown and the crystal exhibited a homogeneous green color without cracks and inclusions of polytypes [36]. The capability of the solution growth process for growth of 3C-SiC crystals, the forgotten polytype, has also been demonstrated recently by using a slightly modified version of the TSSG reactor [37]. This method has produced crystals with reduced stacking fault densities in comparison to 3C-SiC crystals produced by chemical vapor deposition methods.

These studies have clearly positioned the liquid phase growth as a viable route for bulk SiC crystal growth. Nonetheless, there exist many unanswered questions before it becomes a competitor to the standard PVT method. Some of the fundamental questions include the control of dopant concentration and uniformity as well as the incorporation of solvent into the crystal. Finding a suitable crucible material, though graphite shows promise, is a challenge as molten Si is highly corrosive to all current crucible materials. Increased complexity of the growth equipment is also a considerable factor.

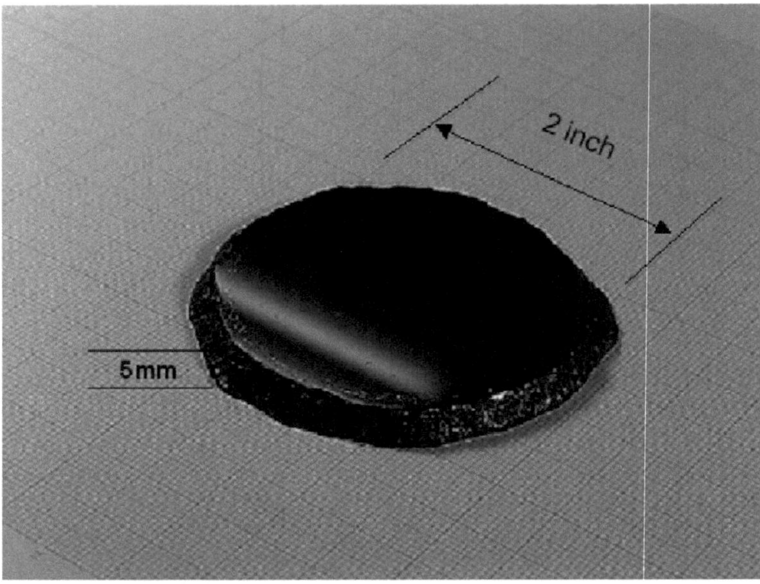

Fig. 2.9 [37] 6H-SiC bulk crystal grown by TSSG technique using Si-Ti-C solution (©Elsevier 2008), reprinted with permission.

2.2.3 Growth Related Issues

2.2.3.1 Polytype Control

A characteristic property of SiC is its ability to exist as over 200 polytypes. The most common polytypes are 3C, 4H, 6H and 15R. As described in Chapter 1, SiC is formed by covalently bonded silicon and carbon atoms in a tetrahedral fashion with each C atom surrounded by four Si atoms and each Si atom is surrounded by four neighboring carbon atoms. These Si-C units are arranged in a hexagonal bilayer with Si and C alternately occupying sub-layers. The staking sequence of theses bilayers along the C-axis determines the polytype. Depending on the terminating atom type, the 0001 face of the SiC has either a C or Si terminating layer.

One of the major obstacles faced by the SiC substrate technology is the different polytype inclusion in crystals. Polytype inclusion during the growth limits the larger diameter single-crystalline SiC substrates. Polytype inclusion also creates nucleation sites for other defects leading to severe quality deterioration [38, 39, 40, 41, 42]. The main reason behind the different polytype inclusion during growth is very low stacking fault energy. This demands exceptional control over thermodynamic and kinetics of the growth process, thus, precise control of thermal conditions and growth pressures are needed [43, 44]. The growth cell must be carefully designed and special attention must be given to the mounting of the seed crystal [14]. In addition to thermal and pressure conditions, the other factors that influence the polytype inclusion include seed surface polarity [21, 11], supersaturation [44, 6, 28, 45], vapor phase stoichiometry [45, 44, 6], impurity levels [46, 47], seed off-cut angle [48, 49], and facet of the crystal [42].

Surface polarity of the seed crystal has a dominant influence of the polytype of the growth crystal in PVT growth. As stated previously, SiC lattice consists of a bilayer in which Si and C making alternating layers. This makes silicon carbide polar in nature, thus, it has two chemically different [0001] crystal surfaces, i.e. the [0001] Si-face and the [0001] C-face, which have different surface energies. It has been shown that the [0001] Si-face has a higher surface energy than the [0001] C-face [11]. The 4H polytype, which has a higher formation enthalpy, always grows on the C-face with the lower surface energy regardless of the polytype of the seed crystal used. Similarly, the 6H polytype, which has a lower formation enthalpy, preferentially grows on the Si-face with the higher surface energy. Many studies have revealed that the polytype of the grown crystal depends on surface polarity rather than the polytype of the seed, thereby indicating the strong influence of the surface energy or surface polarity on the resulting polytype inclusions [50, 11].

Analytical modeling shows a strong correlation between growth temperature and nucleationof different SiC polytypes [51]. For instance, 3C-SiC grows at low temperature while hexagonal polytypes needs high growth temperature due to differences in the energy of formation. 4H-SiC requires growth temperatures lower than that for 6H-SiC growth [43, 44] but 4H and 15R polytypes occur at similar temperature conditions [52]. It is rather difficult to control polytype transition by merely controlling temperature conditions however, as the stacking fault energy is very small. There exists a complex interplay between formation energy and growth condition on polytype inclusion in both the initial stage as well as during the growth. Two parameters that have significant impact on polytype transition are supersaturation and the Si:C ratio in the vapor phase. These parameters are directly controlled by the crucible temperature, temperature gradient, and pressure of the reactor. High supersaturation and low Si:C vapor ratio are crucial to form 4H polytype when grown on the C-face of 6H-SiC. A high axial temperature gradient is needed to meet both these conditions as it allows the use of a high source temperature, thereby producing carbon rich vapor, and low seed temperature, which facilitates supersaturation [6, 45]. When the 4H polytype is grown on the C-face of a 4H-SiC seed, high reproducibility is achieved if growth starts at low supersaturation levels (growth rates of 100 μm/h). However, once the proper growth front has developed, supersaturation level can be increased in order to obtain high growth rate [44]. This is typically achieved by reducing the inert gas pressure while keeping the high temperature gradient to achieve the desired Si:C ratio.

Impurities in source materials also found to affect SiC polytype stabilization. Rare-earth elements such as Sc and Ce tend to stabilize the 4H polytype [47, 22]. The exact role of these elements is still unknown; however, some speculative assumptions have been made. One assumption suggests that these metals may enrich the vapor with C via carbides [14] and others think impurities work as a surfactant which changes the surface energy of the nuclei [50]. Nitrogen, which incorporates into the lattice, also has a significant impact on the polytype stability of 4H-SiC.

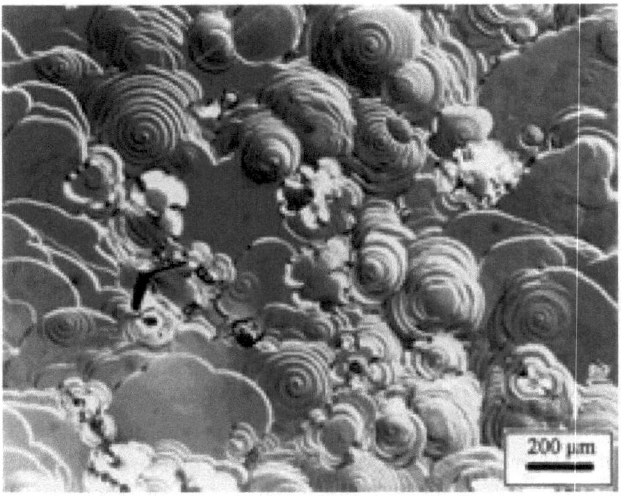

Fig. 2.10 [14] SEM image of growth surface with several growth centers assumed to be formed through spiral growth mechanism around screw dislocations (©Wiley 1997), reprinted with permission.

This effect is attributed to the influence on the Si:C concentration ratio and to the relative enrichment of the growing surface with carbon [47, 50, 46].

2.2.3.2 Substrate Defects Control

Substrate defects are detrimental to the SiC device technology because these defects typically propagate to the subsequent epilayers. The reduction of the substrate defects is perhaps the most critical challenge faced by SiC wafer technology. As most studies centered on reducing defects in wafers grown using standard PVT method, most of the discussion here related to the standard PVT growth of 4H- and 6H-SiC with growth direction parallel to the c-axis, unless otherwise stated.

A distinct feature to PVT growth is the existence of growth spirals (Figure 2.10). Many growth factors such as instabilities in growth parameters and the quality of the seed crystal can lead to secondary and three-dimensional nucleation causing the spiral growth [14]. These spirals have a strong relation to the formation of crystal defects. For instance, these spirals can move across one another as the growth progress resulting in low-angle grain boundaries due to mis-orientation of one spiral with respect to the other. Other major defects due to spiral growth include dislocations, crystal mosaicity (domain structure), and micropipes (open core screw-dislocations).

Among all defects, micropipes are seen as the major threat that can potentially limit the viability of SiC as a commercial semiconducting material. Micropipes, the hollow core of a large screw dislocation, penetrate the entire crystal along the growth direction (when growth is parallel to c-axis) and are replicated to the device epitaxial layer [53]. Therefore, they become detrimental to the device performance. The causes and the formation of micropipes have been widely discussed and there exist many contradicting views and opinions on the mechanisms involved in micropipe formation. Most opinions revolve around Frank's Theory [54] that predicts micropipes are formed on a screw dislocation that possesses a large Burgers vector. Recent studies using synchrotron white-beam x-ray topography (SWBXT) further confirm that micropipes are large Burgers vector screw dislocations and the magnitude of the Burgers vector of a micropipe can range from 2 to 7 times the unit c lattice parameter [55]. Size of the micropipe has a direct relationship to the magnitude of the Burgers vector.

Several possible growth-related sources of micropipe formation have been identified [56, 57]. They can be categorized into three groups: thermodynamics, kinetics, and technological related. The thermodynamic sources can be thermal field uniformity, vapor phase composition, vacancy supersaturation, dislocation formation, and solid-state transformation. The kinetic sources include nucleation processes, growth phase morphology, inhomogeneous supersaturation, and capture of gas bubbles [14]. The technological aspects include process instabilities, seed surface preparation, and contamination of the growth system. Better understanding of these sources along with experimental investigations and accurate modeling of the growth process have resulted in a vast improvement of growth technology and successful control over micropipe formation. Particularly in recent years, there has been a steady progress in reducing of micropipe densities. Currently, four inch n-type 4H-SiC wafer with zero micropipes are commercially available [13].

Micropipe defects are seen in crystals grown by the seeded sublimation growth (standard PVT) with the growth direction parallel to c-axis. Even though the micropipe defects are inherent to seeded sublimation growth, crystals grown using its ancestry methods, namely Acheson and Lely, rarely exhibits any micropipes. The phenomenon is credited to the growth direction as Acheson and Lely crystal growth occurs in the directions perpendicular to the c-axis, that is [1100] and [1120] (a-axis). The micropipe suppression for these off-axis growth methods is attributed to strain relaxation. Strain relaxation largely depends on growth axis and differs significantly between the crystal grown parallel to the c- axis and perpendicular to the c-axis [58]. These finding led to new research directions and many research have been focused on growing SiC perpendicular to c-axis. Results confirm that micropipes can be the eliminated when crystals are grown in the [1100] and [1120] direction using seeded sublimation growth [59, 60]. Although this approach has shown merit in reducing micropipes, at its current stage, the commercial feasibility of this method is rather remote because this method tend to yield a large number of basal plane stacking faults in the grown SiC crystal [58, 61]. Recently, a method called inverted "repeated a-face" growth was introduced as a modification of the perpendicular to c-axis growth process [62]. This method is far superior in terms of its ability to reduce the micropipe density along with the density of dislocations. However, the complexity of this method has prevented its wide spread commercial implementation.

2.2.3.3 Electrical Property Control

Resistivity is one of the most important factors for any semiconductor material. The challenge is to control the residual and intentional doping levels for desired device applications. High power devices, one of the major application areas of SiC, require low resistance substrates in order to reduce power losses caused by parasitics and contact resistances. In contrast, semi-insulating substrates are essential to achieve low dielectric losses and reduced device parasitics for devices and circuits operating at microwave frequencies.

Nitrogen is commonly used as the n-type doping impurity and aluminum is the main p-type dopant for SiC. They create relatively shallow donor and acceptor levels in the SiC bandgap. Recently, phosphorus has been proposed as a replacement for nitrogen as the n-type donor, because the solubility of phosphorus in SiC is higher than that of nitrogen [27]. However, the standard PVT method used for commercial production of SiC substrates uses nitrogen. Nitrogen doping is carried out by incorporating nitrogen gas into the wall of the growth crucible by exploiting the porosity of the graphite. Aluminum, on the other hand, is directly mixed into the SiC source material though aluminum depletion during the process is a key drawback to this approach. This source depletion during the growth has negatively impacted the production of p-type substrates by standard PVT.

The characteristics of dopant incorporation to 6H and 4H polytypes are generally similar. In the case of seed polarity, the doping incorporation varies substantially between the [0001] C-face and [0001] Si-face. Nitrogen incorporation in crystal grown on the [0001] C-face of 6H- or 4H-SiC exhibits higher carrier concentration than crystal grown on the [0001] Si-face under identical growth conditions by a factor of 3 to 5 times [21]. Figure 2.11 shows the dependency of dopant incorporation on nitrogen flow for the n-type doped 6H-SiC grown on the [0001] C-face and [0001] Si-face [21]. A reverse effect has been seen with regards to aluminum incorporation as it prefers the [0001] Si-face. In the case of undoped SiC crystals, the crystals grown on the [0001] C-face show n-type conductivity while the crystals grown on the [0001] Si-face exhibit p-type conductivity. The preferential doping incorporation is attributed to surface kinetic effect as N incorporate to C sites and Al incorporate to Si sties on the crystal.

The surface polarity effect along with precise control of growth parameters has been used for effective control of the impurity levels. Currently, highly doped (10^{20} cm^{-3}) n-type 4H- and 6H-SiC and semi-insulating (10^{14} cm^{-3}) 4H-SiC substrates are commercially available. The lowest reported resistivity values for 4H and 6H-SiC are 0.0028 and 0.0016 Ω-cm, respectively [14], while the highest resistivity value reported is for 4H-SiC: greater than 10^5 Ω-cm [13]. Owing to its intrinsic crystal properties, 4H-SiC has higher carrier mobility with smaller anisotropy com-

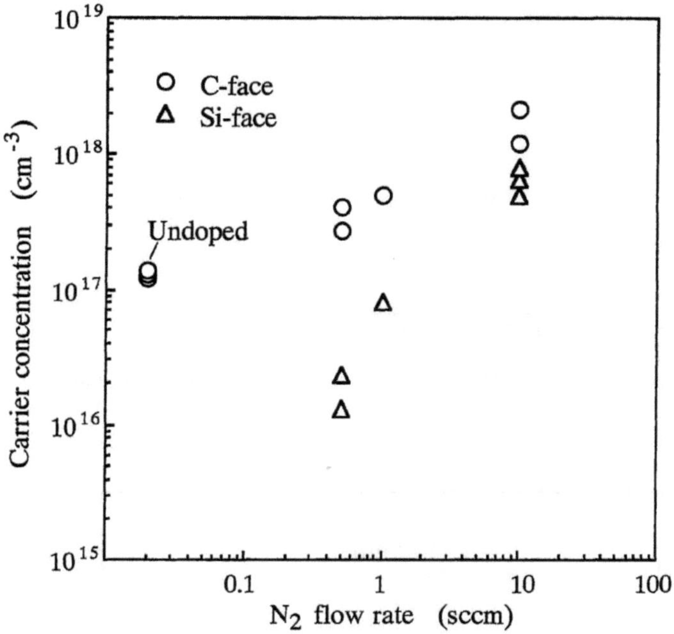

Fig. 2.11 [21] The dependence of the dopant incorporation on nitrogen flow for the n-type doped 6H-SiC grown on [0001] C-face and [0001] Si-face (©Japan Society of Applied Physics 1995), reprinted with permission.

pared to 6H-SiC. These key properties are highly beneficial for high power and high frequency device applications, and due to this particular benefit, the current market trend is leaning towards 4H-SiC substrates.

2.2.4 Current Status

Larger diameter SiC substrates with low defect densities, high crystalline quality, and controlled impurity levels are critical for realizing the full potential of SiC as a mainstay material for electronics, photonics, and microsystems. Continuous relentless research and development efforts from both academia and industry have guided SiC substrate technology to new heights in terms of quality and the size. Figure 2.12 shows the progress of increasing the substrate diameter during the last two decades [63]. Currently 100 mm diameter 4H- and 6H-SiC substrates are commercially available, and it is expected that 150 mm diameter substrates will soon become available.

In addition to increasing the wafer diameter, reduction in defect densities, especially micropipes, took precedence over the last ten years because both performance

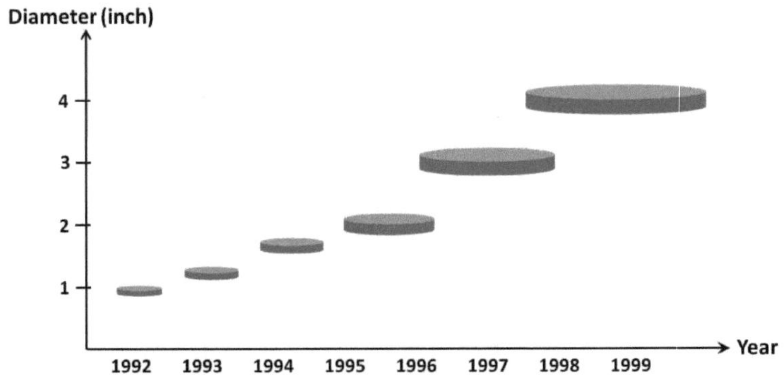

Fig. 2.12 [63] Increase in wafer diameter for 4H-SiC vs. year (©Elsevier 1999), reprinted with permission.

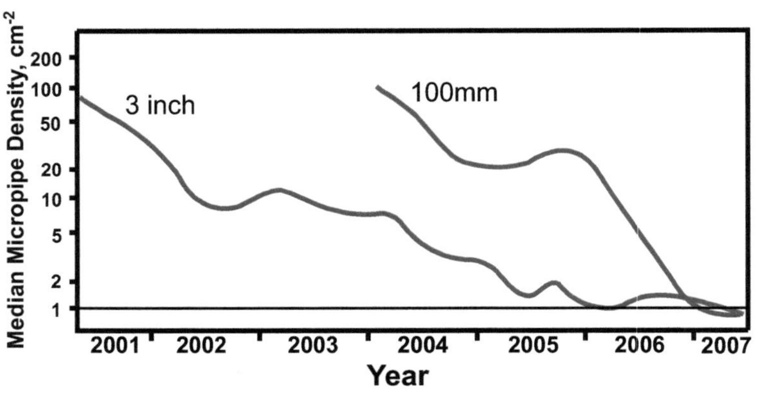

Fig. 2.13 [64] The reduction of median micropipe density on n-type 4H-SiC vs. time for 100 mm and 3 inch wafers (©Trans Tech Publications 2009), reprinted with permission.

and yield largely depend on these defects. Figure 2.13 shows the recent progress in micropipe reduction of PVT grown SiC substrates [64]. It should be noted that this data is related to SiC growth along the c-axis as this is the current method of commercial production of SiC substrates.

It is clear that SiC substrate technology has made a tremendous progress during the last two decades. These advances indisputably provide a solid foundation for the realization of the full SiC microsystems. Further reduction of substrate defects, increase of wafer diameter, and decrease of production cost will lead to the full potential of SiC as the materials for harsh environment microsystems.

2.3 Epitaxial Thin Films

The realization of SiC based harsh environment microsystems is in part determined by the ability to produce SiC electronic devices. For all SiC electronics, epitaxial film growth is a necessity as diffusion doping of substrate is not feasible, and direct ion implantation into the substrate, as is typical for Si, produces inferior electrical quality in SiC [2]. Therefore, the progress and performance of SiC electronics devices to a large extent depends upon the quality, reproducibility, and high volume production capability of epitaxial SiC layers. This includes reducing of defect densities, eliminating polytype inclusions, and controlled doping of n-type and p-type with doping profiles ranging from extremely low (10^{14} atom/cm^3) to very high (10^{20} atom/cm^3). Epitaxial growth of SiC on a variety of substrates has been reported that include homoepitaxial growth on SiC substrate and heteroepitaxial growth on silicon substrate [65, 107, 67]. The discussion here simply focuses on homoepitaxial growth on SiC substrates because it is a prerequisite that devices have to be on SiC substrates for harsh environment compatibility. Furthermore, nearly all SiC high performance devices are currently fabricated using homoepitaxial films as they provide superior electrical characteristics.

Homoepitaxial growth of SiC films can be achieved by various means, each with its own advantages and disadvantages. Selection of a growth technique is in part determined by the application requirements and the technological maturity of the technique. The reported homoepitaxial techniques for SiC can be categorized into vapor phase epitaxy (VPE), liquid phase epitaxy (LPE), and vapor-liquid-solid (VLS) epitaxy. The latter is a recently introduced, novel epitaxial approach which shares the common fundamentals of VLS nanowire and nanotube growth [68, 69]. The following section will briefly discuss SiC epitaxial growth techniques in terms of their maturity and impact as well as advantages and disadvantages.

2.3.1 Vapor Phase Epitaxy (VPE)

Three main techniques are applied in Vapor Phase Epitaxy (VPE), namely chemical vapor deposition (CVD), sublimation epitaxy, and high temperature CVD. Among these, CVD is the most matured and researched technique for epitaxial growth of SiC. It is the core technique adapted by the industry for commercial production of epitaxial SiC (epitaxial thin film on SiC substrate) wafers. Epitaxial SiC wafers are generally referred to as SiC epi wafers.

2.3.1.1 Chemical Vapor Deposition (CVD)

In CVD growth of SiC, carbon- and silicon-containing gaseous compounds are transported to a heated single-crystalline SiC substrate where the homoepitaxial growth occurs through a surface-induced chemical reaction. Depending on the poly-

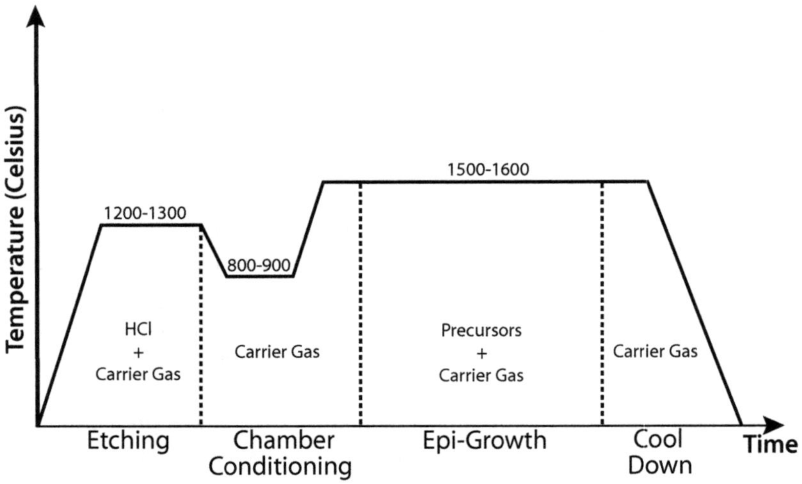

Fig. 2.14 [20] Typical temperature profile vs gas flow conditions for CVD homoepitaxial growth of SiC (©Taylor and Francis 1997), reprinted with permission.

type and the reactor configuration (hot wall or cold wall), the growth temperature can be considerably different, but typically, is above 1200° C. Based on the deposition pressure, CVD can be categorized into atmospheric pressure CVD (APCVD) and low pressure CVD (LPCVD). APCVD was a dominant technique for SiC epi growth throughout the 1980s and early 1990s, mainly due the availability of the APCVD reactors. With advancement in LPCVD, researchers shifted focus to LPCVD as it offers better control of the growth process in terms of gas phase nucleation and impurity levels. Most current industrial processes are now based on LPCVD, yet APCVD is still being used in some research laboratories throughout the world.

There are many different gases used as Si and C precursors. For Si sources, SiH_4, SiH_2Cl_2, $SiCl_4$, and Si_2H_6 are the most popular choices. C_3H_8 is the most common C source. CH_4, C_2H_2, and CCl_4 have also been explored as C precursors. In almost every case, hydrogen is used as a carrier or growth-facilitating gas. Process conditions for a particular gas combination may need to be tuned due to differences in dissociation and nucleation kinetics of the precursors. A typical process for homoepitaxial growth of SiC using SiH_4 and C_3H_8 are shown Figure 2.14 [20].

The process starts with etching the substrate with HCl gas around 1200-1300°C. This helps to atomically clean the surface, which in turn reduces the defects in the epilayer. After etching is stopped, the temperature is reduced and the H_2 flow rate is increased. High hydrogen flow ensures the flushing of residual HCl from the reaction tube. Then reactor temperature is increased again to the appropriate epi growth temperature while keeping the hydrogen flow constant. Once the temperature uniformity is achieved, the precursor gases are introduced into the reaction tube. At the end of crystal growth, the precursor gases are shutoff, the reactor temperature is main-

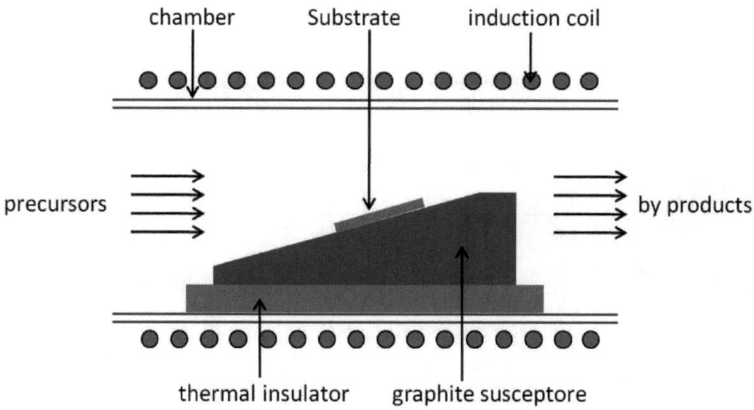

Fig. 2.15 The reactor geometry of the cold-wall SiC epitaxial CVD reactor used in early stage of SiC technology.

tained at the same level for few minutes while hydrogen is flushing out the residual precursor gases before the temperature is decreased. This prevents the inclusion of low temperature polytypes. CVD growth reactors for SiC epi can be categorized into two major groups, cold wall and hot wall. Prior to the mid 1990s, cold-wall reactors were primarily used for SiC epitaxy. Cold-wall reactors were common in III-IV semiconductor processing and adapting that technology for initial SiC development was relatively easy. However, the trend has changed toward hot-wall reactor because of the intrinsic advantages of hot-wall reactor based processes. Currently, hot-wall rectors are the dominant configuration in commercial SiC epi wafer production.

Many reactor configurations have been exploited in both cold-wall and hot-wall CVD. Each of these reactors has its own advantages and limitation in terms of control growth and throughput. Some are highly suited for industrial level production while some offer more flexibility for research and development. Most common reactor types will be discussed below. This will be followed with process related issues and advances in CVD epitaxial growth of SiC.

Early research into SiC epi growth was performed using converted gallium-arsenide rectors similar to one shown in Figure 2.15. The cold-wall configuration is achieved by using a double-walled quartz tube with water circulated between the walls. The wafer is placed on an inductively heated graphite susceptor. To ensure the cold-wall conditions, the susceptor is placed on thermal insulation. The susceptor is slightly tilted with respect to the gas flow to minimize the gas depletion effect for uniform growth. The capabilities of reactor configuration were limited in terms substrate size, temperature uniformity, and growth rate, yet it can be valuable tool for understanding aspects of epitaxial growth.

Later, many cold-wall reactors with slightly different configurations were introduced. Figure 2.16 shows two of the notable reactor configurations used in SiC epitaxy. Figure 2.16(a) a represent the multi-wafer barrel reactor developed by Kong and co-workers [70]. The multiple wafer capability and reduction in particle accu-

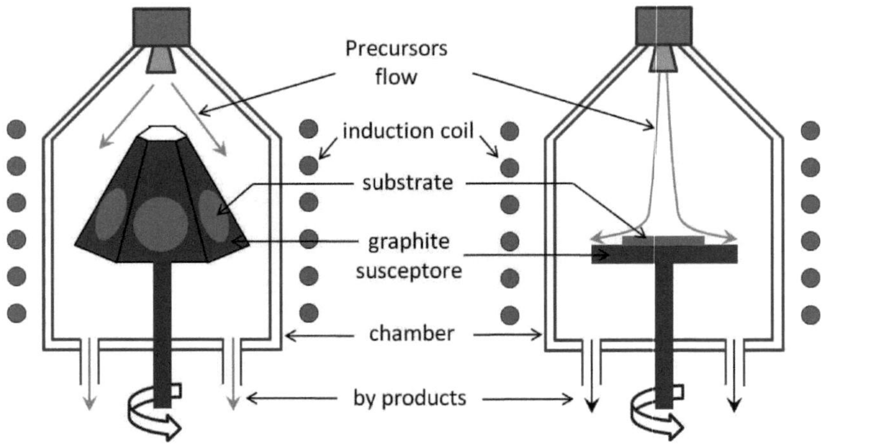

Fig. 2.16 Schematic drawings of cold-wall multi-wafer barrel reactor and rapid rotating reactor.

mulation on the substrate surface is a key advantage to this geometry. This is one of the early era reactor design which still being used in both epitaxial and polycrystalline SiC growth. The rapid rotating vertical reactor (Figure 2.16(b)) was developed in the late 1990s and has shown promises. This reactor would be very good choice for research scale applications; however, further scaling of this reactor to facilitate large area wafers for industrial scale production is challenging. The depletion of source gas and dopant precursor along with temperature nonuniformity causes variations in thickness and dopant profiles [71, 72].

Despite heavily used in early stage SiC development some inherent disadvantages of the cold-wall reactor configuration limits its viability as an industrial tool for SiC epi. Most of the shortcomings are related to the thermal uniformity of the reactor because the area above the substrate is not actively heated. The hot wall reactors enjoy better thermal uniformity in both lateral and vertical directions. In comparison, the temperature gradient in the vertical direction over the substrate surface can be as large as 220K/mm in a cold-wall reactor and that is nearly ten times higher than what is achievable in a hot wall reactor [73].

One of the key disadvantages that relates to the temperature profile of the cold-wall reactor is poor precursor dissociation efficiency, which directly translates into the growth rate. The maximum growth rate of cold-wall reactors is around 5 μm/h [74] while the growth rate of a hot-wall reactor can be as high as 100 μm/h [75]. However, most hot-wall reactors operate with growth rates below 25 μm/h to control the quality of the epitaxial layer [76]. In addition to the growth rate related issues, the vertical temperature gradient over the substrate leads to excessive Si supersaturation in the gas phase, causing nonuniformity in the epilayer [77]. In the hot-wall configuration, gas phase Si aggregation is minimized due to high temperature surroundings [78]. Furthermore, hot-wall reactors provide long-term stability of the growth environment, and continuous growth over 30 hours is possible with-

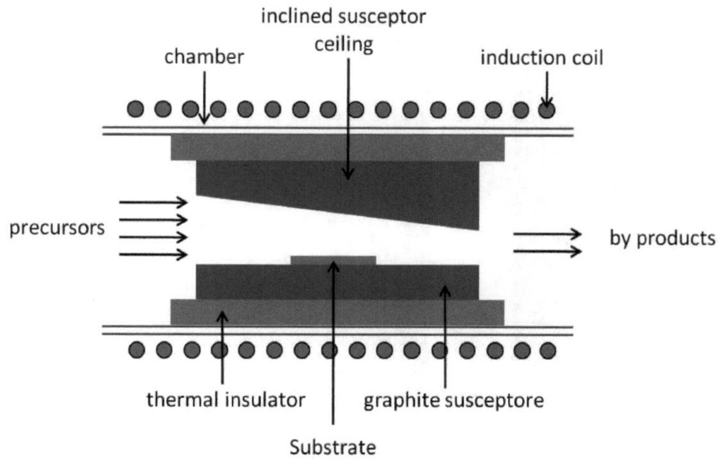

Fig. 2.17 Conceptual diagram of the horizontal hot-wall CVD reactor for epitaxial SiC.

out significant degradation of the SiC growth front [79]. These decisive advantages attracted many users to adapt to hot-wall reactor technology. Some of the widely used hot-wall reactors are discussed blow.

The hot-wall reactor concept was first introduced by Kordina *et al.* in 1994 [80]. This was a horizontal geometry reactor module. Later, the reactor was further improved for highly uniform epitaxial layer growth [81]. Figure 2.17 conceptually represents the horizontal hot-wall CVD reactor. The graphite susceptor used here is a rectangular-shaped hole which runs along the entire length of the reactor with inclined ceiling. This geometry of the susceptor increases the gas velocity and reduces the depletion effect in order to grow a uniform epitaxial layer. The susceptor is encircled by thermal insulator which is placed inside an air cooled quartz tube. The thermal insulator reduces the heat loss due to radiation and consequently, hot-wall reactors consume less power (20-40 kW) than cold-wall reactors. The thermal insulator also helps maintain thermal uniformity. To date, this reactor geometry is one of the most widely used hot-wall reactor configurations.

Another hot-wall configuration called a chimney reactor has received considerable interest due its ability to achieve high growth rates [76]. This is a vertical reactor similar to one shown schematically in Figure 2.18. It likewise has a hollow susceptor with an internal rectangular cross-section. The symmetrical nature of the susceptor provides symmetric temperature and gas flow distributions allowing the substrates to be mounted on opposing sides of the inner walls. Typically, the gas flow is upward through the reactor. The flow is facilitated by free convection due to the high temperature process. The growth rate of this reactor can be as high as 50 μm/ hour. Since the high growth rate introduces more defects in the deposited film, the reactor is generally operated below 30 μm/h to produce high quality epilayers.

The introduction of the multi wafer hot-wall planetary reactor concept can be considered a significant step towards high throughput production of SiC epitaxial

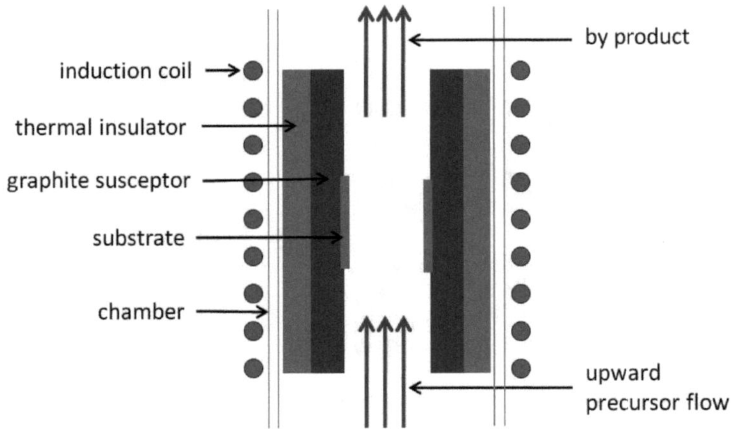

Fig. 2.18 [76] Schematic diagram of the chimney CVD reactor for epitaxial SiC ((©Elsevier 2002), reprinted with permission.

wafers. The planetary reactor concept was originally developed by Frijlink *et al.* in the late 1980s for the growth of III-V compound semiconductors [82]. This design went through a few iterations [71, 83] before becoming current high throughput industrial scale SiC epi hot-wall reactor. The schematic illustration of the reactor concept is displayed in Figure 2.19. The precursors enter from the top at the center of the reactor and flow outward radially. The gas transport properties of this configuration result in a decrease in growth rate as the susceptor radius increases. The growth rate decreases because of precursor depletion as well as an increase in the boundary layer thickness due a rapid drop in gas velocity as the area increases. In this reactor configuration, this effect is successfully eliminated by the rotation of the individual wafers about their individual axes. Experimental evidence confirms achieving extremely good thickness uniformity (1.5%) and dopant uniformity (6%) over 100 mm substrates [84]. Extremely high wafer to wafer uniformity is also reported. The growth rate is typically around 10 μm/h.

A very promising hot-wall reactor concept was introduced recently for high growth rate and uniformity [85]. The highest growth rate reported in this method is 250 μm/h. Optimal epilayer conditions are achieved at lower growth rates, typically around 80 μm/h. Thickness uniformity of 1.1% and dopant uniformity of 6.7% over 100 mm substrate have been reported for this method. The reactor is essentially an improved version of the cold-wall rapid-rotation reactor. The reactor concept is schematically shown in Figure 2.20. The uniform growth is achieved by controlling the gas flow and the temperature distribution in the reactor. First, precursor gases are introduced from an off-centered inlet and the susceptor moves up and down during the growth, resulting in changes in gas flow distribution. The vertical position RF coil can also change independent of the susceptor position to control the vertical thermal distribution. Despite a promising start, the complex nature of the reactor hardware and operation has hindered the wide spread usage of this method.

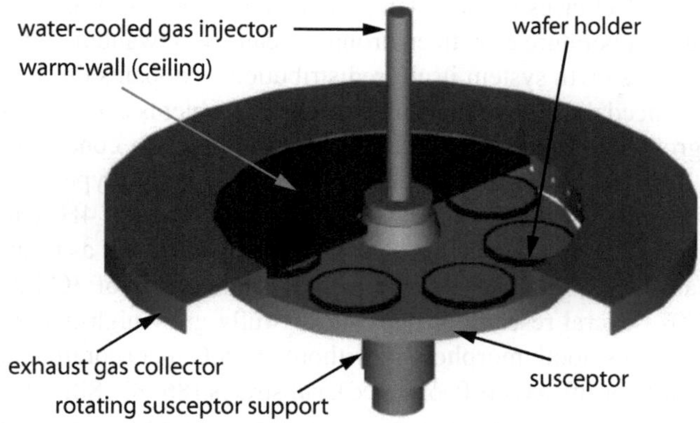

Fig. 2.19 [84] Schematic representation of the multi-wafer epitaxial SiC hot-wall reactor that uses the planetary rotation principle (©Trans. Tech. Publishing 2009), reprinted with permission.

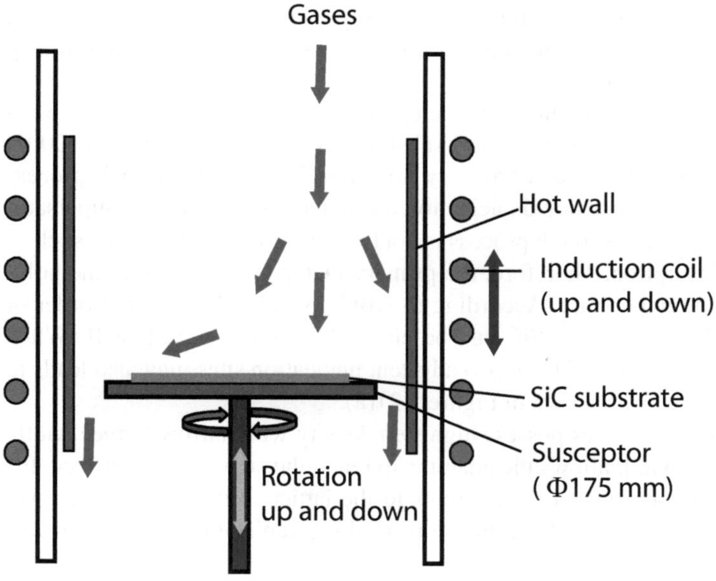

Fig. 2.20 [85] Schematic representation of epitaxial SiC reactor that uses up and down motion of the susceptor and off-centered gas delivery for optimal uniformity (©JSAP 2008), reprinted with permission.

In addition to quality and throughput improvements by novel CVD reactor designs and process optimization, the growth aspect of CVD has also been heavily studied. Two major breakthroughs in growth of SiC CVD epilayers occurred in late 1980s and early 1990s, namely step-controlled epitaxy and site competition epitaxy. These advances have had a tremendous impact on the SiC electronics industry.

Step-controlled Epitaxy: In general, growth of homoepitaxial 6H-SiC layers performed on well-oriented [0001] faces requires temperatures of 1700-1800 °C. Unfortunately, this very high temperature growth environment causes: unwanted impurity contamination from the growth system itself, redistribution of dopants through diffusion, and thermal-induced damage to the epilayer. These problems can be mitigate by lowering the growth temperature but low temperature growth conditions results in inclusion of 3C-SiC, the low temperature polytype. This polytype mixing due to 3C-SiC inclusion is a serious issue in CVD epitaxial growth of 4H- and 6H-SiC polytypes. The 3C-SiC inclusion in hexagonal epilayers is known as triangular defects as they can be distinctly identify by the triangular shape of 3C-SiC crystals. In the late 1980s, several research groups successfully grew high-quality homoepitaxial 6H-SiC with a smooth morphology without 3C-SiC incorporation at 1400-1500 °C using vicinal, or off-axis (off-oriented), substrates [86, 87, 88]. Surface steps existing on the off-oriented substrates serve as a template for replication of the underlying polytype. This technique of growing epilayers on off-axis substrates is known as step-controlled epitaxy. This technique was a significant breakthrough in homoepitaxial growth of SiC as it enabled production of device-quality epilayers with replication of substrate polytype at reduced growth temperature (>300 °C). This is beneficial in reducing contamination from the reactor wall and minimizing unwanted dopant diffusion.

Figure 2.21 schematically illustrates the epitaxial growth process on (a) a well-oriented and (b) an off-oriented 6H-SiC substrate [89]. The well oriented [0001] face consists of vast terraces and has very low step density. The growth process proceeds through two-dimensional nucleation on the terraces due to high supersaturation on the surface. The growth process is controlled by surface reactions such as adsorption and desorption. Therefore, the primary factor that determines the polytype is the growth temperature. According to ABC notation, the stacking order of 6H-SiC is ABCACB while 3C-SiC can be either ABCABC or ACBACB. When 3C-SiC grow on well-oriented face, two adjacent nucleation sites may also leads to double positioned twins as shown in Figure 2.21(a).

The off-oriented substrates possess high step density with narrow terrace width. The smaller terrace width allows the adatoms to reach the step through surface diffusion and the incorporation of the adatoms to the lattice at the step. The growth process is governed by bonds from the step resulting replication of the substrate polytype.

The initial studies on step-controlled epitaxial were performed on 6H-SiC polytype; however, later studies shows its viability to homoepitaxial growth of other polytypes including 4H-SiC [22]. The remarkable success in step control growth led to a burst in research to understand the growth mechanism and the factors affecting step-controlled growth. The growth process of step-controlled growth is found to be mass transport limited, not surface reaction limited. Therefore, supersaturation conditions should be controlled to promote mass transport limited growth and to prevent two-dimensional nucleation. Growth rate, growth temperature, and terrace width all influence the growth mechanism [89].

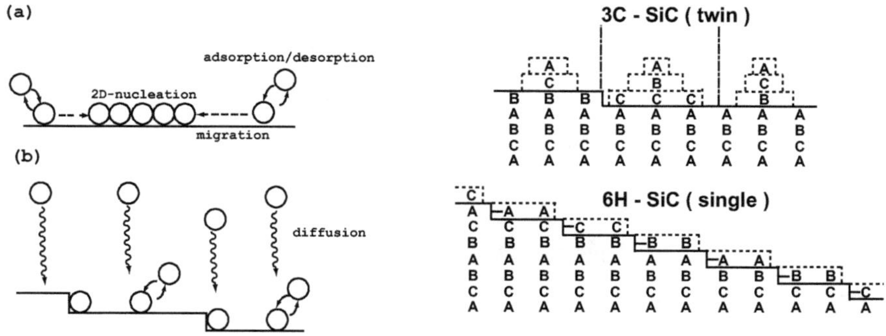

Fig. 2.21 [89] Schematic representation of the growth mechanism on (a) well-oriented and (b) off-oriented of 6H-SiC [0001] faces (©AIP 1994), reprinted with permission.

As the off-angle of the substrate increases, the terrace width decreases. Figure 2.22 gives the experimental data of the growth rates for various off-angles of the [0001] face of 6H-SiC substrate at 1500 °C [90]. It clearly shows 3C-SiC growth on both the Si-face and C-face of well oriented substrates at 1500 °C while homoepitaxial growth is observed on off-oriented substrates under the same experimental conditions. It is also clear that off-angle orientation as small as 1 degree can induce the step flow growth conditions. The data further reveals that surface polarity does not affect growth on off-oriented substrates. As stated in Section 2.2.3.1, 4H polytype, preferentially grow on C-face whereas 6H polytype preferentially grows on the Si-face. The nonexistence of the polarity effect on step controlled growth is attributed to the relatively low activation energy for the step controlled growth (3 kcal/mol) in comparison to that of well-oriented substrate (C-face 20 kcal/mol, Si-face 22 kcal/mol).

Site Competition Epitaxy: The precise control of dopant incorporation is essential to fully realize the intrinsic advantages of SiC for high power, high temperature, and high frequency electronics. This was a difficult task, especially in obtaining lightly-doped material because of unintentional doping that was difficult to prevent at the high deposition temperatures. The introduction of site competition epitaxy by Larkin *et al.* in early 1990s [91] provided a solution. In this method, the controlled doping is based on adjusting the C:Si ratio within the growth reactor. Earlier research pointed out that dopant atoms occupy specific sites of the SiC lattice, specifically nitrogen occupies the carbon site while aluminum occupies the silicon site [92, 93]. High C:Si ratio results in an increase in carbon concentration in the growth environment forcing a competition between nitrogen and carbon for the C-sites on active growing surface of SiC lattice. A similar situation occurs when the C:Si ratio decreases, which results in the relative increase in silicon concentration in the growth environment forcing a competition between Al and Si for Si sites on the active growing surface of the SiC lattice. Therefore, controlling C:Si ratio of the feeding gasses can be used for controlling dopant atom incorporation. This is the basis for "site-competition epitaxy."

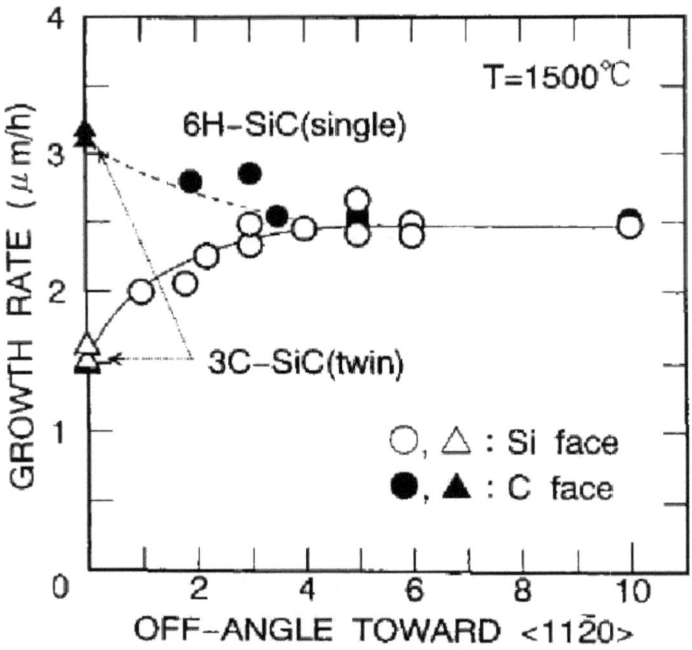

Fig. 2.22 [90] The effect of the off-angle of the substrate on the growth rate and polytype of a epitaxial SiC layer grown on 6H-SiC at 1500 °C with flow rates of SiH_4 and C_3H_8 are 0.30 and 0.20 sccm, respectively. Triangles represent 3C-SiC and circles represent 6H-SiC ((©)AIP 1993), reprinted with permission.

Site competition epitaxy is used not only for N and Al incorporation, but also has been worked for common impurities such as boron and phosphorus [94]. This technique has been heavily implemented to control impurity levels for both intentional and unintentional dopant incorporation during the epitaxial CVD growth of 6H-, 3C-, 15R-, and 4H-SiC. For instance, when the C:Si ratio was increased from 2.3 to 10 for undoped epitaxial growth on the Si face of 6H-SiC, the layers changed from n-type to p-type due to suppression of N incorporation. This technique enables production of both p-type and n-type epilayers with carrier concentration down to 10^{14} cm^{-3}. These types of semi-insulting substrates with low carrier concentrations are desirable for high power SiC devices. Highly doped (up to 10^{20} cm^{-3}) epilayers have also been achieved by site competition epitaxy, benefitting device technologies that require low parasitic resistances.

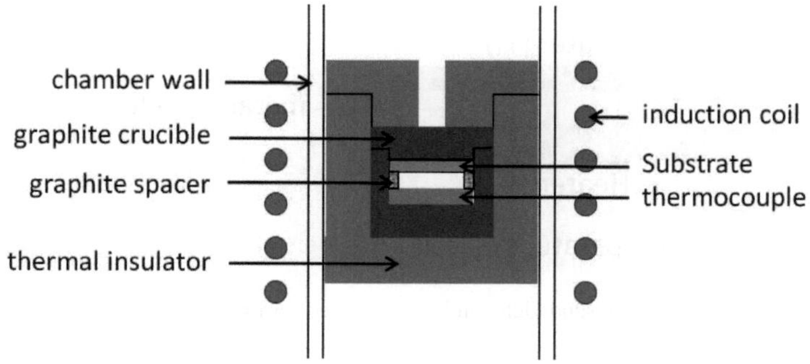

Fig. 2.23 [95] Schematic diagram of the reactor used in close space technique for epitaxial SiC growth (©Elsevier 1999), reprinted with permission.

2.3.1.2 Sublimation Epitaxy (SE)

Sublimation epitaxy (SE), sometimes referred as close space technique, involves a very similar process to the standard PVT method (modified Lely method) described previously. The SiC substrate is placed very close to the source material; the typical distance between source and substrate is around 1 mm, compared to 20 mm for standard PVT. The reactor is schematically depicted in Figure 2.23 [95]. The process is carried out at slightly lower temperatures (1800-2200 °C) and higher pressure (up to 1 atm) than standard PVT. The biggest advantage of SE over previously described CVD techniques is the capability of growing epitaxial films at very high rates — as much as 1000 μm/h has been demonstrated [95]. One drawback to this method is that changing dopant levels or dopant type during the growth is not possible.

2.3.1.3 High-Temperature Chemical Vapor Deposition (HTCVD)

This method was initially introduced by Kordina *et al.* It shows greater promises in terms of growth rate, purity, and dopant control [23, 81]. High-Temperature Chemical Vapor Deposition (HTCVD) is suitable for both SiC epitaxy and bulk growth. A brief description on the HTCVD apparatus and the process can be found in section 2.2.1. As stated earlier, the growth process in HTCVD occurs through sublimation of gas phase nucleated Si_xC_y clusters; it greatly differs from the conventional CVD processes. The process temperature is extremely high (1800-2300 °C) and helium is used as the carrier gas to prevent etching of the susceptor by hydrogen. Growth rates as high as 800 μm/h have been reported, and it is comparable to that of boule growth by the standard PVT method. Along with HTCVD crystal growth, this method for producing epilayer has been adapted in industrial environment [16].

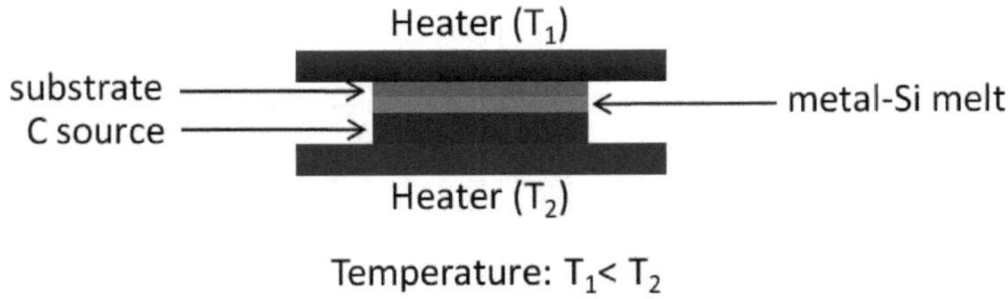

Fig. 2.24 Conceptual drawing of the sandwich configuration used for liquid phase epitaxial growth of SiC.

2.3.2 Liquid-Phase Epitaxy (LPE)

LPE was one of the widely used techniques for SiC epitaxial layer growth for device applications during the 1990s [96]. Although CVD is currently the preferred method for SiC epitaxial layer growth, recently there has been a renewed interest in LPE due to its ability to reduce the micropipe defect density. As mentioned in section 2.2.2, the growth process occurs at low temperature in equilibrium conditions and micropipe generation is not energetically favorable under these conditions. The epitaxial growth process is mainly carried out using the traveling solvent method [97, 98], in contrast to liquid phase bulk growth that primarily utilizes the top seed solution growth method. Figure 2.24 schematically represents the reactor geometry. A temperature gradient is maintained between the source and substrate to facilitate the growth process. The growth rate of LPE can be as high as 300 μm/h. Both n- and p-type doping can be achieved by selecting proper source materials. A key limitation to this method is the inability to switch doping level and conductivity type during the growth [79].

2.3.3 Vapor-Liquid-Solid Epitaxy (VLS)

VLS is comparatively a new technique for growth of epitaxial SiC [99]. The growth mechanism is based on the well known VLS process of SiC nanowire or whisker growth [69]. The reactor configuration for VLS SiC epitaxial growth is depicted in Figure 2.25. A silicon melt is formed in a crucible by a combination of metal and silicon powders. There are few steps involves in the growth process, which starts with transporting of the carbon precursor (C_3H_8) to the surface of the liquid where precursor dissociation and dissolution of carbon in the silicon melt take place. Then, carbon transport from the vapor-liquid interface to the liquid-solid interface occurs where crystallization of SiC happens. The wetting properties of the substrate surface by Si melt and the height liquid droplet on the substrate is crucial for uniform growth rate. The crucible was designed in such a way that uniform droplet height is maintained to mitigate growth rate variations as shown in Figure 2.25.

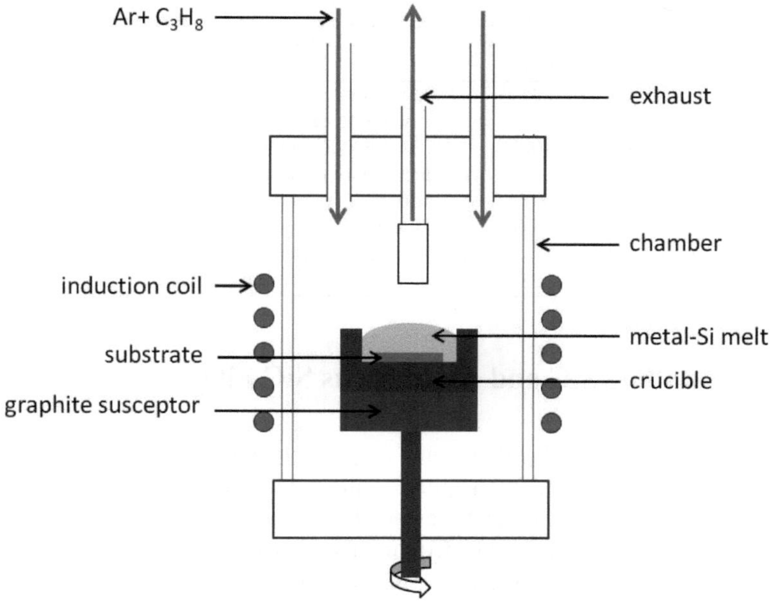

Fig. 2.25 [99] Schematic illustration of VLS growth reactor for epitaxial SiC growth ((©RSC 2004), reprinted with permission.

The maximum growth rate obtained with this method is around 35 μm/h. The VLS concept is still at the research stage. The commercial implementation of this method is yet to be seen.

2.3.4 Current Status of SiC Epitaxial Wafers

Table 2.2 summarizes the key features of the SiC epitaxial growth techniques discussed above. Some of the techniques are already being used in the commercial production environments while others are still at the research stage. Most industrial production of SiC epitaxial films produces 4H- and 6H-SiC polytypes. Both n- and p-type doping are available with a wide range of carrier concentrations (9 x 10^{14} to 1 x $10^{19}/cm^3$) and doping uniformity better than 10% over a 4 inch substrate. Thicknesses up to 50 microns are routinely produced with thickness uniformity better that 2% [13, 100, 101]. Despite commercial availability and wide spread usage of epitaxial SiC layers, there still remain many challenges to eliminate defects in epilayers. For example, micropipe formation due to substrate imperfections is yet to be solved. Even though 3C-SiC inclusion is suppressed to an acceptable limit by step controlled growth, further improvement in this area is desirable.

Table 2.2 Key features of the current SiC epitaxial growth techniques

Technique	Growth Temp. (°C)	Growth rate (μm/h)	Comments	Ref.
CVD	1400-1500	25	Adopted in commercial environments, Growth rate can be as high as 100μm/h	[76]
SE	1800-2000	1000	Changing doping level and switching conductivity type during growth not possible	[95]
HTCVD	1800-2300	800	Adopted in commercial environments	[16]
LPE	1700-1800	300	Changing doping level and switching conductivity type during growth not possible	[97]
VLS	1500-1600	35	At initial research stage	[99]

2.4 Polycrystalline SiC and Amorphous SiC Films

SiC MEMS technology is today dominated by poly-crystalline 3C-SiC (Poly-SiC). Poly-SiC growth does not require an epitaxial alignment and consequently the high temperature pre-carbonization can be avoided. This allows the growth of poly-SiC on variety of substrate materials. Common substrate materials include single-crystalline silicon and single-crystalline SiC as well as silicon dioxide, silicon nitride, and polysilicon thin films [102, 103]. The ability to grow on substrates other than SiC offers great flexibility in device fabrication using poly-SiC. The ability to deposit at lower temperatures minimizes the temperature mismatch between SiC structural layers and the underlying sacrificial or isolation layers. From a fabrication flexibility perspective, it enhances the ability of achieving complex MEMS structures through surface-micromachining as poly-SiC can deposit on a wide range of sacrificial and isolation materials [106]. The low temperature deposition also permits the use of the poly-SiC layer as a protective coating for MEMS devices fabricated using other materials such as polysilicon [104, 105].

Initially, APCVD reactors used for epitaxial SiC growth were adapted for poly-SiC deposition [107, 108]. Later, as the process matured, various reactor configurations as well as wide range of deposition techniques were introduced for poly-SiC. LPCVD is currently the leading deposition technology for poly-SiC. However, other techniques are also used to produce poly-SiC including PECVD, magnetron sputtering, ion-beam sputtering, and ion implantation. Each of these techniques has its own advantages and disadvantages with regards to control of electrical and mechanical properties and deposition characteristics. Later in this chapter, each of these techniques will be detailed briefly.

Amorphous SiC (a-SiC) retains most of the chemical and mechanical properties of poly-SiC and has been used for many MEMS structures. In particular, the ability to deposit at very low temperatures is attractive from an integration point of view. All the techniques used for poly-SiC deposition can also be used to deposit a-SiC, and depending on the technique, the deposition temperature can be as low as 25 °C. This even allows use of polymeric layers or heat sensitive materials as sacrificial or isolation layers. Furthermore, it can also be used as a coating material to enhance the chemical resistance of MEMS made using other materials. Currently, PECVD is the dominant technique for producing a-SiC; however, newly emerging techniques such as ion-beam assisted deposition show promise with regards to deposition characteristics such as line-of-sight deposition and no hydrogen in the deposited film.

The remainder of this section will detail various deposition techniques mentioned above and outline advantages and disadvantages of each.

2.4.1 APCVD

APCVD was the primary technique used for deposition of poly-SiC during the early phase of SiC MEMS technology. APCVD was the natural choice at that time because it was widely used in epitaxial growth of SiC films for electronics. So extending that knowledge to poly-SiC deposition was relatively easy. Most APCVD poly growth is carried out using reactors that were originally used for 3C-SiC heteroepitaxial growth on Si [107, 108, 109]. Both vertical and horizontal reactors are used. The growth can be accomplished either by dual precursors such as silane and propane as the Si- and C-containing precursor gases [108] or by using a single precursor such as hexamethyldisilane (HMDS) [109, 110]. In both cases H_2 is used as the carrier gas, and stoichiometric poly-SiC films on Si wafers are achieved at temperatures above 1050 °C.

At the initial phase of SiC MEMS, APCVD played a critical role in demonstrating the suitability of poly-SiC thin film technology for harsh environment MEMS applications. It was successfully used to produce MEMS-grade poly-SiC films. Nonetheless, commercial implementation of APCVD for SiC MEMS is hindered by some inherent disadvantages with regards to scaling up the process volume and lowering growth temperature. APCVD SiC reactors use inductively heated graphite susceptors similar to those shown in section 2.3.1.1. This restricts the substrate size and reactor load, limiting throughput. The deposition temperature is over 1050 °C, which is not suitable for depositing poly-SiC on wide range of materials. Furthermore, the high deposition temperature is also a concern for monolithic integration with ICs. In order to address these issues, LPCVD and PECVD methods were proposed.

2.4.2 LPCVD

LPCVD is a well established technique in silicon semiconductor and MEMS processing. This technique offers precise control of gas transport properties compared to APCVD, permitting deposition of thin films with excellent uniformity and extremely good step coverage. These characteristics are highly critical factors for MEMS fabrication. Moreover, excellent purity, large-size wafer capability, and multi-wafer capacity of LPCVD reactors are attractive for high throughput production of high quality films.

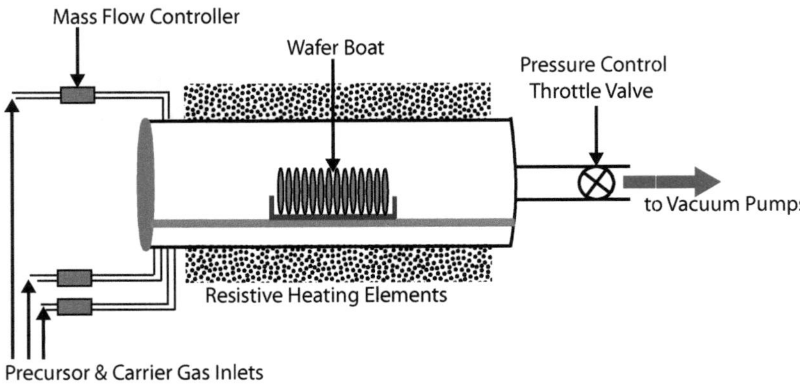

Fig. 2.26 Schematic diagram of a high throughput large diameter hot-wall LPCVD reactor.

Over many years, a large number of feasibility studies, done in small scale reactors, have proven the usefulness of LPCVD as a method for poly-SiC thin film growth [112, 113, 114, 115, 116, 117]. Those studies have shown that LPCVD offers the flexibility to grow poly-SiC films that have a wide range of electrical, mechanical, and chemical characteristics. Encouraged by these initial results, recent efforts have been focused on developing process for poly-SiC deposition in large area multi-wafer reactors. Zorman *et al.* were the first to report depositing poly-SiC films on 100 mm diameter silicon substrates in a reactor capable of handling 100 wafers at a time [118]. From then on, a tremendous amount of research has been performed to optimize deposition parameters and increase throughput. Today, LPCVD processes are available for substrates as large as 150 mm in diameter. The optimization of LPCVD processes to improve film properties for specialized applications continues to be an active area of research.

Figure 2.26 is a schematic illustration of a hot-wall LPCVD reactor used in poly-SiC growth. The reactor is resistively heated as opposed to inductive heating in APCVD. Nitrogen and hydrogen are commonly used as carrier gases. The reactors are very similar to the reactors used in poly-Si processes, except that the temperature of poly-SiC reactors can go as high as 1200 °C. The deposition temperature varies with process conditions and precursor type.

A wide array of single and dual precursors has been exploited for poly-SiC. Single precursor includes disilabutane (DSB), trimethylsilane, and hexamethyldisilazane (HMDS). As for the dual precursor, silane and dichlorosilane are commonly used as the silicon precursor while methane, butane, and acetylene are used as carbon precursor. As the process optimization is based on precursor type, the following discussion is mostly limited to precursors that have reasonably well developed processes for poly-SiC based MEMS production.

For single precursor poly-SiC growth, 1,3 disilabutane (DSB) is the most explored to date. It is capable of producing high quality poly-SiC around 800 °C and can be used to deposit a-SiC at temperatures as low as 650 °C [116, 117]. One of the notable distinctions for DSB is the high growth rate, which can be as high as 55nm/min at 800 °C [116]. There is an extensive set of data available with regards to process conditions as well as the electrical and mechanical properties of the poly-SiC films deposited using DSB [119, 120, 121, 122]. For instance, in-situ nitrogen doped films show resistivity as low as 0.02 Ω-cm [120] and carrier concentrations up to 6.8 x 10^{17} cm^{-3} [123]. One of the key deposition characteristics of DSB-based LPCVD is the highly conformal nature. The low temperature deposition along with highly conformal deposition characteristics is highly desirable for wear and chemical resistive coatings for released MEMS structures. Some examples include wear resistive coating on microscale engine components fabricated using silicon [124] and a chemically resistive coating of a silicon double ended tuning fork resonator[125] and capacitive strain gauge [126]. As for a structural material for MEMS, poly-SiC deposited using DSB has been used for many functional MEMS structures such as Lamé mode filters [127].

Single source precursors such as methylsilane [103], trimethylsilane, and HMDS [109] have similar potential as a source material for LPCVD poly-SiC. Methylsilane has been recently used in a large diameter multi-wafer format LPCVD rector [128]. Key advantages of using methylsilane as a single precursor over DSB are that it is readily available and is substantially less expensive. In terms of deposition temperature, methylsilane can also produce poly-SiC at 800 °C. Initial results show great promise with regards to electrical and mechanical property control. Nitrogen doped films made using methylsilane exhibit resistivities comparable to DSB deposited films [129]. Based on current status, methylsilane is a sound single precursor that needs attention from poly-SiC developers.

The combination of dichlorosilane (SiH_2Cl_2) and acetylene (C_2H_2) dominates dual precursor based LPCVD poly-SiC deposition because of its early start and well characterized processes for MEMS grade films. The deposition temperature for MEMS grade poly-SiC using this gas chemistry is 900 °C. Starting with a feasibility study by Wang *et. al*, this gas combination was quickly adapted to poly-SiC deposition on large area (100-150 mm) wafers in multi-wafer LPCVD reactors [118]. From then on, LPCVD deposition with SiH_2Cl_2 and C_2H_2 has played a crucial role in SiC MEMS development. Extensive research has been performed on this precursor combination for many years, so there exists a vast knowledge base on process conditions and corresponding material properties. Outstanding stress control over a wide range from high tensile to medium compressive, including crossing zero stress, has been achieved. In terms of n type doping, the nitrogen atomic concentration up to of 2.6 x $10^{20}/cm^3$ and resistivity down to 0.02 Ω-cm have been reported [130]. In terms of technological maturity, the process using this dual precursor system is ready for wide spread use. Most of the MEMS devices discussed in Chapter 4 have been fabricated using thin films deposited with this technique. High-temperature, shock-resistance strain gauge [131], harsh environment accelerometer [132], and pressure sensor for in-cylinder pressure measurements [133] are but a few.

Despite new PECVD and other physical vapor deposition methods emerging for creating poly-SiC films, to date LPCVD has become the gold standard technique for poly-SiC MEMS mainly because of the relentless push towards achieving high quality films with reproducible electrical and mechanical properties. In the next section, process parameters that are used for controlling highly critical factors such as doping, residual stress, and strain gradient of LPCVD poly-SiC films will be discussed.

2.4.2.1 Doping of LPCVD Poly-SiC Films

For many MEMS applications, poly-SiC with tunable levels of conductivity are either highly desired or required. As mentioned in Chapter 1, diffusion doping is not an option for SiC, and the controlled inclusion of impurity atoms is accomplished through the addition of a dopant precursor to the reactor during the growth. With regards to MEMS applications, the conductivity takes precedence over the impurity type (n or p). Therefore, in most cases, n-type doping with nitrogen is preferred for poly-SiC due to processing ease. Furthermore, nitrogen has the shallowest ionization energy in SiC comparison to other dopant atoms such as phosphorus, aluminum, and boron [1].

Doping of epitaxial as well as bulk growth of SiC are routinely done using gaseous nitrogen and ammonia [134, 135]. However, dopant incorporation in MEMS-grade poly-SiC is relatively difficult due to low temperature deposition requirements along with the required material characteristics of the poly-SiC. First, the doping precursor must decompose at low temperatures. The existence of grain boundaries in poly-SiC also complicates dopant incorporation and carrier transport at low temperature. The trapping of dopant atoms in grain boundaries and reduction of adatom mobility directly affects the growth process, resulting in changes in the film properties [119, 136, 137].

There were many concurrent studies on doping low-temperature-deposited poly-SiC using ammonia. The first successful attempt was reported by Wijesundara *et al.* for doping of poly-SiC grown by DSB at 850 °C [119]. Resistivity down to 0.02 Ω-cm was achieved. Later, doping of poly-SiC at growth temperature as low as 800 °C was demonstrated confirming the of practicality of NH_3 as a doping precursor for low temperature deposition of poly-SiC [124]. Figure 2.27 shows the resistivity changes due to increase in ammonia in the feed gas. Currently, NH_3 is being used as the dopant precursor for poly-SiC deposited from both single and dual precursor deposition systems [130, 129].

The key challenge in doping poly-SiC is to optimize electrical properties while keeping the desired mechanical properties for MEMS applications. It is shown that doping can lead to changes in the growth characteristics such as growth rate, degree of crystallinity, grain size, and residual stress as well as influences the Young's Modulus of poly-SiC. Changes in the degree of crystalline quality due to doping variations was initially observed by Wijesundara *et al.* [119]. This was further confirmed by Zhang *et al.* showing a decrease in the lattice constant as doping concentration increases [123]. Furthermore, recent studies show that changes in grain size due to doping influences the quality factor of flexural-mode poly-crystalline silicon carbide (SiC) lateral resonators [136]. These results shows that mechanical properties and the doping level of poly-SiC are intertwined, and it is important to optimize both

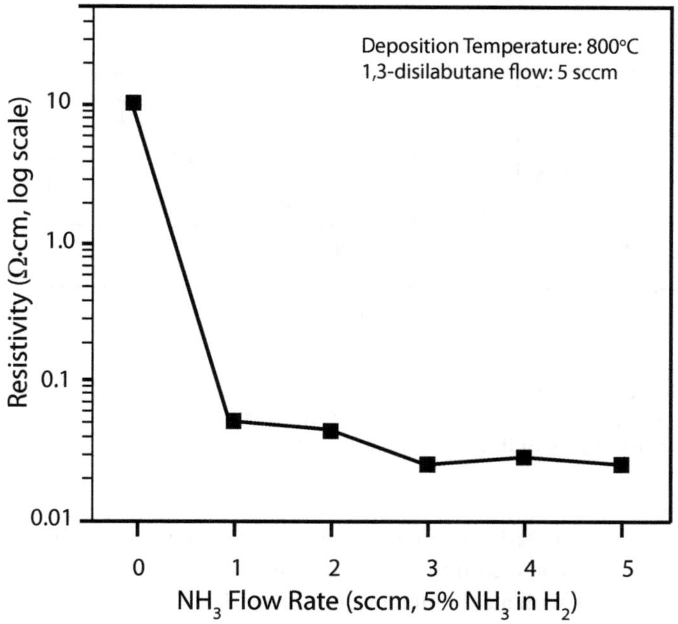

Fig. 2.27 [124] The resistivity variation of poly-SiC films deposited at 800 °C as a function of the NH_3 (dopant precursor) flow rate (©Elsevier 2003), reprinted with permission.

electrical and mechanical properties together for the realization of poly-SiC MEMS devices. These studies also suggest that doping can be used as a tool for tailoring device characteristics.

2.4.2.2 Residual Stress and Stress Gradient Control on LPCVD poly-SiC

As mentioned at the beginning of the chapter, the residual stress and stress gradient are two highly critical factors that have to be addressed at the materials level in order to realize MEMS devices. Processes have to be developed to reduce both the average stress and stress gradient such that they are sufficiently small to be useful in creating free-standing microstructures that do not break, buckle, or curve out of plane. The control of these two parameters in poly-SiC thin films is very challenging due to several reasons. First, deposition is done on top of different materials with varying thermal properties. The thermal mismatch and elevated deposition temperatures cause thermal induced stress at room temperature. Lowering the deposition temperature and selecting substrates and interlayer materials with similar thermal expansion can be helpful. However, for LPCVD deposition, the temperature is primarily dictated by the precursor used. The material selection is governed by the poly-SiC surface micromachining technology. Currently, polysilicon, silicon dioxide, and silicon nitride are the primary set of materials used in poly-SiC micromachining. Therefore,

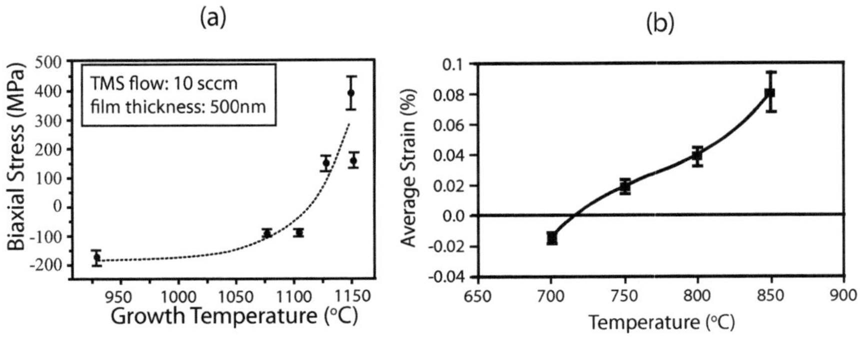

Fig. 2.28 [113, 121] Influence of deposition temperature on the average stress of LPCVD poly-SiC thin films deposited using (left) Tetramethylsilane (TMS) (©AIP 2000) and (right) 1-3 disilabutane (DSB) showing zero-crossing (©SPIE 2003), reprinted with permission.

the material choice is rather limited. The other factor which highly influences the stress and stress gradient of deposited film is the microstructure of the film. Typically, poly-SiC growth occurs with a columnar or grain structure. The size of grains or columns has a strong correlation to residual stress. Furthermore, the grain or column size vary as the growth progresses, resulting in variation in stress through the film thickness, causing strain gradient.

The deposition temperature has been one parameter explored for controlling both residual stress and stress gradient as it has a direct relation to the microstructural properties of the films including grain, column, and crystallinity. Figure 2.28 displays the temperature dependence of poly-SiC films deposited by (a) tetramethylsilane and (b) DSB [113, 121]. Temperature can be used to change thin film stress from highly compressive to medium compressive to tensile, including crossing the zero stress point. However, temperature alone cannot be used to optimize MEMS films as it also changes other parameters such as resistivity and crystallinity.

One notable development in controlling the film stress is reported for dual precursor poly-SiC deposition is to control deposition pressure [138]. Pressure influences the gas transport property of the reactor thereby causing microstructural changes, specifically grain or column size. In this case, low stress poly-SiC films were obtained while keeping the stoichiometry, crystallinity, and resistivity at optimal conditions. Figure 2.29 shows the residual stress versus pressure for films deposited at 900 °C using SiH_2Cl_2 and C_2H_2.

Other aspects of changing deposition parameters have also been investigated. Roper *et al.* reports that changes in the C:Si ratio has an effect on the stress levels of poly-SiC deposited using DSB [139]. Adding small amounts of SiH_2Cl_2 to a reactor during deposition changes the elemental composition and grain size of the resulting film, thereby changing the stress level.

The cause for a variation in stress through the thickness of a material is mainly attributed to changes in the stress level of the films as a function of microstructure variation along the thickness. Structural changes have been observed for both

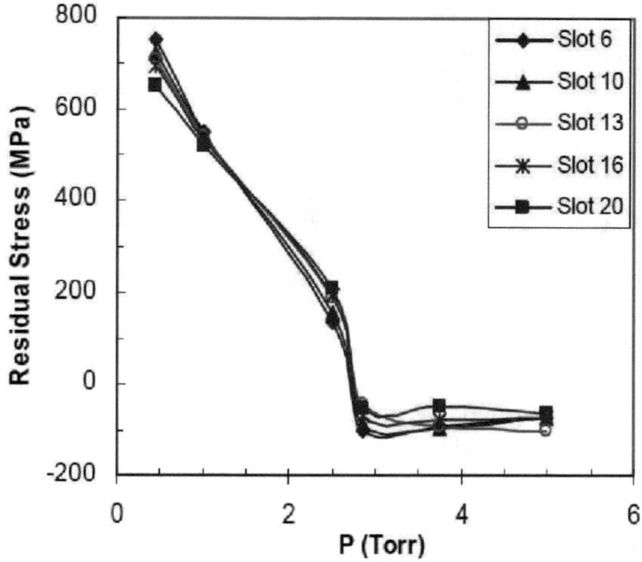

Fig. 2.29 [138] Residual stress as a function of deposition pressure for poly-SiC films deposited at 900 °C using SiH_2Cl_2 and C_2H_2 (5% in H_2) (©Trans Tech Publications 2004), reprinted with permission.

columnar and grain growth of poly-SiC as the thickness increases. Figure 2.30 is a TEM image of poly-SiC interface grown on a SiO_2 substrate [110]. It clearly indicates the column size varies as the thickness increases. For grain growth, a similar behavior was observed by Fu *et al.* for poly-SiC grown on a SiC substrate [140]. One successful approach to suppress this structural variation is by disrupting the growth process. Deposition can be done in few successive runs in order to stop the continuation of underline grains or columns. Another way of addressing this issue is layer by layer deposition with each different layer having different doping levels to further modulate the stress. More detail on this process can be found elsewhere [141].

Even though there is room for further optimization and development, LPCVD poly-SiC has matured to a level that makes it useful for many MEMS applications. At this stage, the majority of poly-SiC MEMS are still limited to academic research institutions (Figure 2.31). The transition of academic poly-SiC processes for MEMS to commercial production of SiC microsystems is gated by unavailability of supporting SiC electronics.

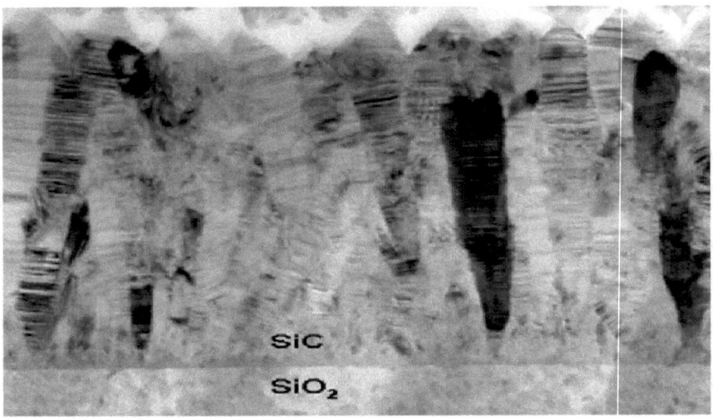

Fig. 2.30 [110] TEM image of poly-SiC grown on SiO_2 showing the variation of the size with film thickness (©Elsevier 2009), reprinted with permission.

2.4.3 PECVD

As a structural material for harsh environment MEMS, PECVD deposited amorphous SiC (a-SiC) exhibits the desired chemical and mechanical properties [143]. PECVD deposition of a-SiC has drawn significant interest because of its very low deposition temperatures (<600 °C). This enables integration of a-SiC into a wide range of substrates, making PECVD an important technique for SiC microsystem fabrication.

The main areas that may heavily benefit from the use of PECVD a-SiC include device encapsulation, protective coatings, and forming dielectric layers. As with the deposition techniques described earlier, the film properties largely depend on the deposition conditions. Depending on the specific application and maximum thermal budget allowed, deposition parameters can be used to tailor film properties.

In general, SiC PECVD is performed in conventional PECVD reactors with heated substrate holders [144]. Methane and silane are commonly used precursors [144, 145, 146]. It is also possible to use a single precursor such as DSB or methylsilane. The film properties strongly depend on the process parameters including deposition temperature, pressure, plasma power, and gas phase composition. In general, PECVD films exhibits compressive stress but process conditions can be optimized to tailor the stress levels to desired values. Typically, low plasma power and high temperature conditions yield low stress films. Sometimes post-deposition annealing at temperatures between 450 and 600 °C may still be required to reduce stress levels [143].

PECVD deposition of a-SiC can be used as a scaffolding layer as well as the final sealing layer for MEMS devices (see Chapter 5). Early studies demonstrated PECVD a-SiC for encapsulation of membranes. Figure 2.32 shows a SEM image

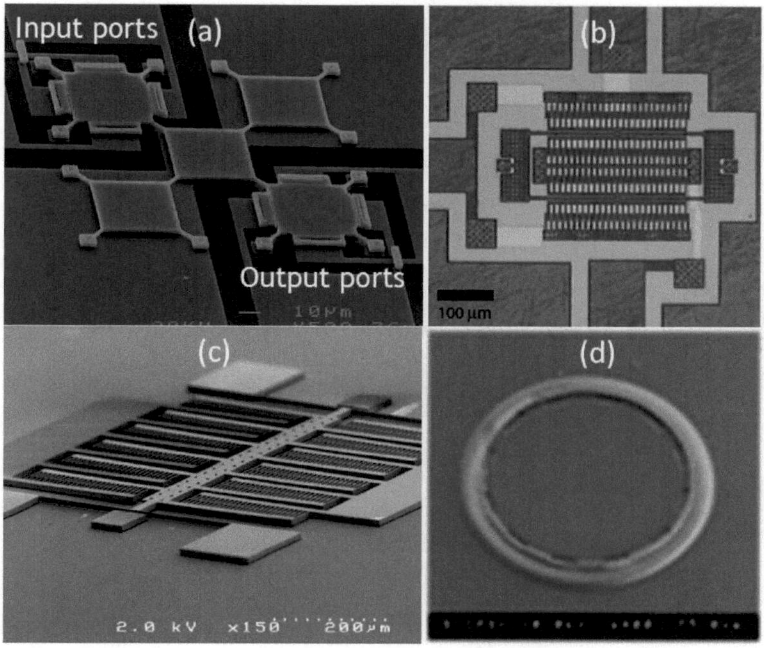

Fig. 2.31 Some of the notable poly-SiC MEMS devices fabricated at various research institutes: (a) a checkerboard filter fabricated using a array of Lamé mode resonators with center frequency of 175 MHz [127], (b) double-ended tuning fork resonant strain gauge that works at 500 °C and survive at 64,000 g shock load [131], capacitive accelerometer for range of 5,000 g [132], and pressure sensor that operates at 575 °C [133]. (©IEEE 2005, SPIE 2009, Trans Tech Publications 2009, Elsevier 2008), reprinted with permission.

of the top-view of a cleaved circular membrane created by PECVD a-SiC encapsulation [147]. These examples demonstrated the potential of PECVD deposited a-SiC. However, as deposited, a-SiC is not conductive enough to become a standalone MEMS material.

2.4.4 Ion Beam Assisted Deposition

Ion beam assisted deposition (IBAD) utilizes physical wafer deposition combined with concurrent ion bombardment to create thin films. The physical wafer deposition technique can be either sputtering or evaporation. Figure 2.33 shows a schematic diagram of an IBAD system equipped with a sputtering ion source (deposition source) and an assist source (ion bombardment source). The role of the assist source can vary depending on the application needs. In some instances, it can be used to densify the deposited film or to create a modified interface between substrate and deposited film. In both cases, ions from a noble gas such as argon are used. This mode of

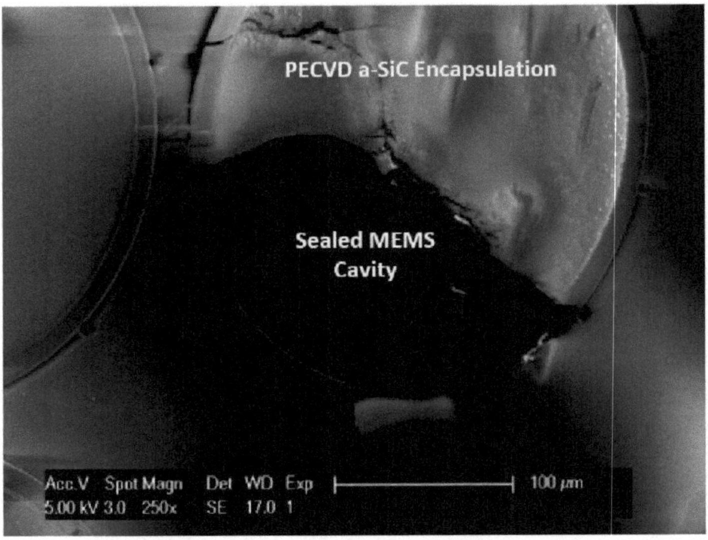

Fig. 4. Top-view of a cleaved circular PECVD a-SiC encapsulation

Fig. 2.32 [147] Top view of cleaved encapsulation membrane fabricated using PECVD a-SiC (©IEEE 2009), reprinted with permission.

operation is suited for depositing single element layers, such as metals. However, if the assist source provide reactive ions to the growing surface, it creates compound materials. In this case, typically, a metal is evaporated or sputtered while the growth surface is continuously bombarded with reactive ions, such as oxygen or nitrogen ions to create metal oxides and nitrides. Currently, IBAD thin films are widely used as magnetic thin films, protective coatings, and hard coatings [148].

To form compound films such as SiC using ion beam sputtering, typically dual ion beam systems are used with separate targets for each type of element [149]. Multiple targets are used because ion beam sputtering is not well suited for sputtering compound material targets such as SiC due to preferential sputtering of one element over the other [150]. That leads to the formation of nonstoichiometric films. However, it has been shown that if a SiC target is sputtered with ion energy around 1200 eV, the sputtering yields 1:1 ratio of silicon to carbon [151]. Encouraged by this observation, an IBAD system was developed for low temperature deposition of a-SiC [152]. This system (Figure 2.33) consists of two ion guns, a target holder, and a substrate holder. The ion energy range of the sputter gun is between 500 and 1500 eV and argon is used as the sputtering gas. The assist ion gun ranges from 50-500 eV. The role of the assist gun in SiC MEMS thin film deposition is for stoichiometric fine tuning of the films as well as strain gradient control of the deposited film [153]. The typical deposition rate of this type of system varies from 5 to 15 nm/min and is a function of the ion fluence of the sputtering source.

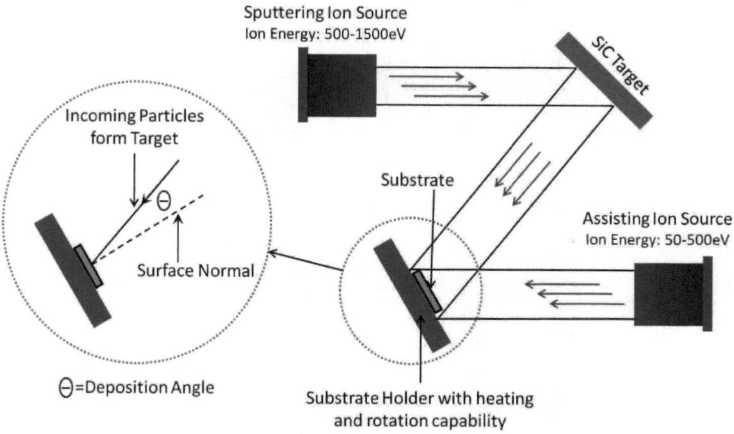

Fig. 2.33 Schematic representation of ion assisted SiC deposition system with definition of the deposition angle.

Films can be created at room temperature or at high temperature with a heated substrate holder. The film stress is compressive in nature and the stress levels decreases as the temperature increases. The deposition pressure also has an effect on stress levels. Typically, higher deposition pressure reduces compressive stress. Deposition angle, θ, also has a huge impact on the stress levels. As θ increases, the compressive stress decreases due to changes in packing density. Assisting ions also increases the compressive stress due to film densification. Therefore, it is desired to do deposition without assisting ions, unless demanded for stoichiometric fine tuning or strain gradient control.

A unique characteristic to ion beam sputter deposition is the high directionality of sputtered particles coming out of the target. This is due to the use of collimated ion beam for sputtering. Therefore, the IBAD technique allows line-of sight deposition, which is highly advantageous for etch hole sealing during encapsulation in comparison to other physical and chemical vapor deposition techniques.

Typical zero level encapsulation schemes deposit a sealing layer on a porous scaffold film after releasing the underlying MEMS device (see Chapter 5) [154]. Currently, CVD is the main technique employed for sealing the scaffolding layer [154]. In some instances, PECVD and material reflow methods have also been investigated [132, 155]. Despite being a widely used technique, CVD tends to mass load the released MEMS devices due to the conformal nature of the deposition [156]. IBAD enables etch hole sealing without unwanted mass loading of released MEMS structures because of its inherent line-of-sight deposition capability. The preliminary results of IBAD for etch hole sealing highlights its potential. Figure 2.34 represent the deposition characteristics of IBAD based etch hole sealing. Results clearly indicate no visible slide-wall or down-hole deposition of SiC.

IBAD for a-SiC is still at an early stage of development. As shown in Figure 2.34, it has a tremendous potential for wafer level encapsulation [152]. Since the deposi-

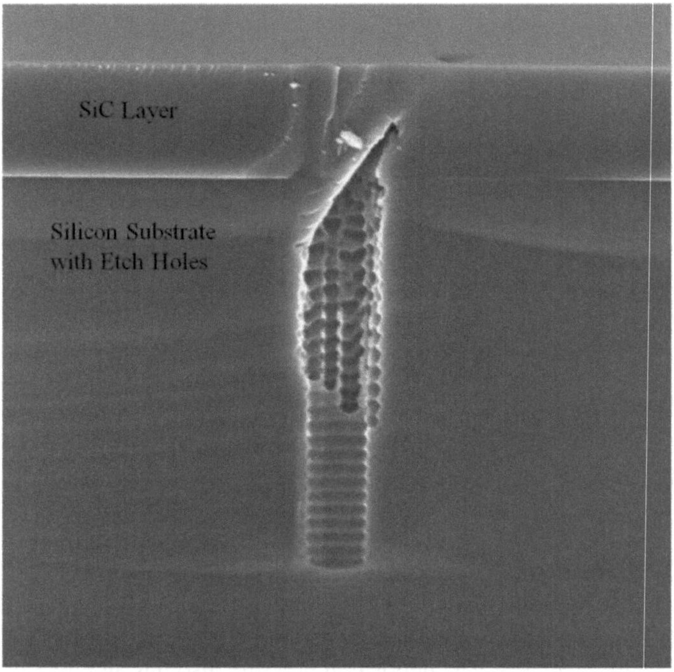

Fig. 2.34 [152] Cross-sectional SEM image of a sealed etch hole with deposition angle of 50 degrees showing no visible deposition on sidewalls and the bottom of the substrate ((©IEEE 2007), reprinted with permission.

tion is typically carried out at pressures on the order of 10^{-6} Torr, this method would be highly attractive for vacuum encapsulation as the low cavity pressure drastically reduces air damping of vibrating MEMS devices. Furthermore, line-of-sight deposition can be used for creating unique MEMS devices such as 3D microstructures by using shadow mask techniques. Despite the lower deposition temperature of an IBAD a-SiC film, it does not contain hydrogen as opposed to PECVD. Hydrogen-free films are highly attractive for high temperature applications due to its high thermal stability.

2.4.5 Other Deposition Methods

Poly-SiC and a-SiC can be deposited using wide variety of techniques. Magnetron sputtering is an attractive technique for SiC since it is a low temperature deposition method very similar to PECVD. However, studies have yet to be focused on utilizing this technique for MEMS grade poly- or amorphous-SiC. Some limited studies on the deposition of a-SiC demonstrates that it is possible to achieve stoichiometric SiC by magnetron sputtering [157]. Exploring this technique as an alternative to PECVD

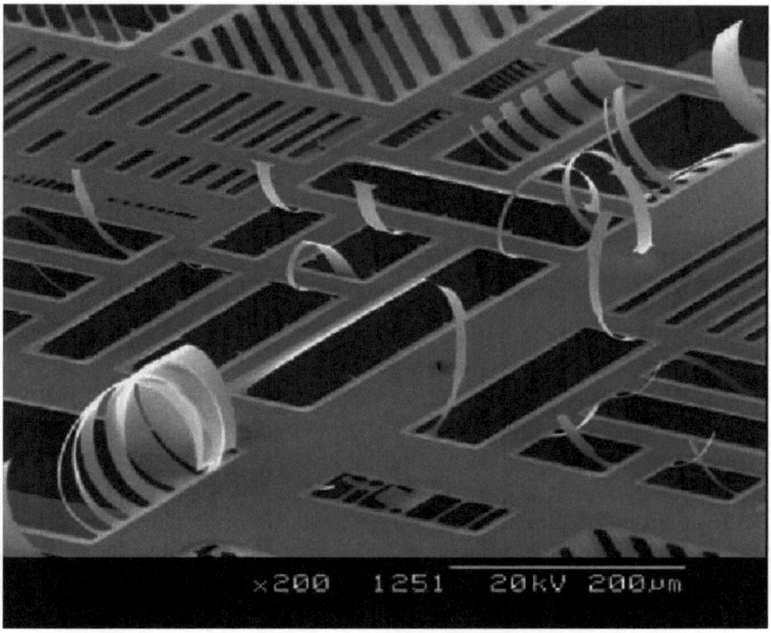

Fig. 2.35 [158] SEM topograph of a released MEMS test structure fabricated using C ion implantation of polysilicon (©IOP 2000), reprinted with permission.

would be beneficial because it enables producing hydrogen free a-SiC films at lower temperature. Further studies need to be focused on controlling of mechanical and electrical properties of the SiC films.

Serre *et al.* reported direct synthesis of SiC microstructures by implanting carbon ions through a shadow mask into a SiC substrate [158]. This is a very attractive process as it allows creation of MEMS devices without etching SiC films. When SOI wafers are used, the top Si layer is converted to SiC by ion implantation, and the underlying oxide layer can be used as sacrificial layer. Figure 2.35 displays a SEM image of a fabricated microstructure using this method. As evident by the SEM images, stress and stress gradient control of the films is an issue. The implantation leads to dislocating atoms in close proximity and densification of the film leading to increased film stress. Further research in this area is needed to mitigate stress factors and understand thickness limitations of this method.

At the current stage of SiC MEMS technology, CVD SiC deposition techniques are mature enough to supports fabrication of complex microstructures that are similar to silicon MEMS. However, the advancement of SiC MEMS demands new deposition techniques that will enhance the prospect of achieving an all SiC microsystem. More specifically, it is highly critical to examine low temperature deposition processes because it allows fabrication of MEMS devices on top of electronics without violating the overall thermal budget of the electronic circuitry.

2.5 Patterning of SiC

The creation of microstructural features by selectively removing SiC films or bulk materials is essential for each device layer of the microsystem. However, high chemical inertness, owing to high bond strength between the silicon and carbon, makes sculpting of SiC quite difficult. This imposes and extra burden on selecting a patterning method, masking materials, and sacrificial layers in order to create microdevices.

Etching, molding, milling, and ablation can be used for micropatterning of SiC materials. As for etching, both dry and wet chemical etch processes have been exploited. Molding of microstructural elements is seen as a way to overcome the patterning difficulties that arise from the high chemical inertness of SiC. Ion milling techniques, such as FIB (focused ion beam), provide opportunities for creating high precision submicron features; however, the serial nature of the process limits the throughput and complexity of structures. Laser ablation of SiC is faster in comparison to FIB but its ability of making microstructures with fine geometric control has yet to be demonstrated. The following section will review the most commonly used methods for selective pattering of SiC thin films and bulk materials.

2.5.1 Dry Etching of SiC

Dry etching of SiC is mainly achieved through plasma-based reactive ion etching (RIE). This method is the primary method used to selectively etch SiC for device fabrication today. RIE dry etching provides precise control of linewidth, sidewall, and surface profile. The removal of SiC in RIE occurs through a combination of physical and chemical processes. Physical sputtering occurs through bombardment of the surface with energetic particles ejected from the plasma. Chemical etching occurs through reaction of active chemical species present in the plasma with surface species. Controlling of these two competing processes are essential for creating an etch profile with the desired side-wall angle and surface finish. Therefore, the selection of reactor type, the process conditions, and gas chemistries are extremely critical.

Various etch chemistries have been developed for plasma-based SiC etching. Most processes have been centered on fluorine based chemistries. Fluorinated compounds such as CHF_3, $CBrF_3$, CF_4, SF_6, and NF_3 mixed with O_2 have been successfully implemented [159, 160, 161]. However, SF_6 and O_2 is the most researched gas combination for Si etching, widely deployed in both research and industrial environments [159]. Hence, it is an obvious choice to explore for SiC etching. Some of the probable chemical reactions associated with SiC etching in a plasma containing fluorine and oxygen is given in equation (2.4), (2.5) and (2.6) [159].

$$Si + xF \rightarrow SiF_x \tag{2.4}$$

$$C + yF \rightarrow CF_y \tag{2.5}$$

$$C + zO \rightarrow CO_z \tag{2.6}$$

The presence of oxygen in a fluorinated plasma facilitate the SiC etch process in several ways. From Eq. (2.6), it is apparent that it is directly involved in the removal of C from SiC. Furthermore, the atomic oxygen in the plasma reacts with unsaturated fluoride species to generate reactive F atoms. Therefore, the oxygen presence in the plasma provides more reactive species for removing C and Si while simultaneously consuming the polymer-forming fluorocarbon species [159, 162].

Plasma etching of SiC with SF_6 and O_2 typically uses metal masks to achieve a high selectivity; however, metal masks are known to create micromasking problem. Micromasking occurs when metal atoms of the mask material are sputtered by the plasma and redeposited in the etch field and act as local mask for the etch process. This results in grass-like structures on the etch surface. In addition, using a metal mask material typically requires a dedicated metal-contaminated tool or extensive cleaning between process changes as metal particles tend to be contaminants for many other processes including CMOS. To eliminate metals and to use widely accepted CMOS compatible masking materials such as SiO_2, and Si_3N_4, chlorine- and bromine-based chemistries have been investigated [163, 164]. Despite achieving reasonable selectivity for SiO_2, and Si_3N_4, chlorine- and bromine-based etching has significantly slower etch rates.

The selection of proper reactor type and operating conditions are also critical for realizing higher etch rate, etch selectivity, vertical sidewalls, and smooth surfaces. For example, conventional reactive ion etching (RIE) systems that use two parallel plates and a RF plasma generator have low plasma densities and high energy species. Therefore, etching of SiC in these type of reactors are dominated by physical sputtering, causing rough etch surfaces and low selectivity towards commonly used masking materials [159]. To overcome these shortcomings, most SiC etching is done using high-density, low-pressure plasma etchers, such as electron cyclotron resonance (ECR), transformer couple plasma (TCP), helicon plasma, and inductively coupled plasma (ICP) reactors [159, 164, 165, 166]. Among these, ICP is the most widely used technique. It offers many advantages over other methods including scalability and low operating cost. The capability of producing high plasma density at lower pressures is highly attractive because it reduces ion scattering, resulting in a significant reduction in lateral etch rate for high anisotropy [159, 166].

From the microsystem perspective, the required etch depth varies with the device type. For example, the etch depth is typically submicron for most SiC electronics applications, whereas it is often one to tens of microns for MEMS, and hundreds of microns for through-wafer holes for some high frequency SiC electronics [165]. To address these varying needs, different masking materials and process conditions have been developed. Photoresist, metal, SiO_2, Si_xN_y are the typical masking materials for selective etching of SiC. To date, the highest etch selectivity of 100:1 is reported for a Ni mask used with a SF_6 and O_2 gas chemistry in an ICP plasma etcher [167]. The reported etch rate is 1.5-1.6 μm/min. This would be an ideal process for many SiC MEMS processes with regards to both etch rate and selectivity. The etch rate can be further increased up to 2.6 μm/min for bulk etching of SiC; however, the etch selectivity decreased to 45:1. The highest etch selectivity for a non-metal masking material, such as SiO_2 and Si_xN_y, is reported by Gao *et al.* using a HBr based gas chemistry in a TCP system [164]. The reported value for etch rate ratio for SiC:SiO_2 and SiC:Si_xN_y is up to 20:1 and 22:1, respectively. While the etch selectivity is reasonable, one of the drawbacks to this etch chemistry is the extremely low etch rate: $<$150 nm/min. Although etch rate can be slightly increase with plasma conditions, the etch selectivity of non-metal mask become poor, leading to fast mask erosion. This results in undesired side-wall angles and rougher etch surfaces [164].

2.5.2 Wet Etching of SiC

Despite the fact that SiC is very difficult to be etched by wet chemical etching, many wet chemical etch processes have been investigated. Some attractive features of wet etching include selective etching of amorphous SiC over single crystalline SiC and dopant selective etching. In addition, wet etching does not require very expensive apparatus.

Wet etching processes for SiC fall into two general categories, namely chemical etching and electrochemical etching. Regardless of the category, the wet etching of SiC involves oxidation of the SiC surface and subsequent dissolution of the resulting oxides. Since the etch process occurs though surface oxidation, many parameters including surface defects, microstructure, dopant type, and seed polarity all can affect the etch characteristics. This section provides a brief overview of wet etching techniques used in SiC. An excellent review of SiC wet etching was conducted by Zhuang *et al.* [170].

Various etch chemistries ranging from highly basic to highly acidic have been investigated for chemical etching of SiC. Molten potassium hydroxide (KOH) etches SiC and the high etch rate is highly desirable for bulk-micromachining of substrates. However, the usefulness of this technique is limited by the lack of control over etching in the lateral direction. Furthermore, the high reactivity of molten KOH towards many other materials presents process compatibility issues. As a low temperature wet etch chemistry, a hydrofluoric (HF) and nitric acid (HNO_3) mixture has been shown to etch a-SiC and poly-SiC [171, 172]. The etch process can be done at room temperature, which is highly desirable. The inherent disadvantage to this method includes poor control over lateral dimensions. Similar to molten KOH, the etch solution is highly reactive to almost every standard masking, sacrificial, and isolation layer material (Si, SiO_2, Si_xN_y, Al, Ti, Ni, Ta, Cr, Mo, W, and Cu) used in traditional microfabrication. Single-crystalline SiC and diamond are highly resistant to HF+HNO_3, so they can be used as etch masks. The etch resistance to single-crystalline SiC over amorphous SiC can be employed to pattern single-crystalline SiC by amorphization of of selective regions. The amorphization of single-crystalline SiC is typically done by ion implantation. Alok *et al.* has demonstrated a highly anisotropic etch by producing a trench with a depth of 0.3-0.8 μm on a 6H-SiC by ion implantation and HF+HNO_3 etching [171].

The electrochemical etch process of SiC occurs via oxidation reactions that forms volatile and dissolvable compounds. Diluted HF is commonly used as the electrolyte in which water serves as the oxidant and HF reacts with SiO_2 formed by the oxidation. The SiC surface act as the anode and generally Pt is used as the cathode. The widely accepted anodic reactions are shown in Eq. 2.7 and 2.8 [170]:

$$SiC + 2H_2O + 4h^+ \rightarrow SiO + CO + 4H^+ \tag{2.7}$$

$$SiC + 4H_2O + 8h^+ \rightarrow SiO_2 + CO_2 + 8H^+ \tag{2.8}$$

where h^+ represents holes.

According to the above reactions, the oxidation reaction is facilitated by hole generation. Ultraviolet (UV) illumination during the electrochemical etch process increases hole generation resulting in an increase in etch rate. This is known as photoelectrochemical etching (PEC) [170]. The feasibility of employing of PEC etching for SiC has been demonstrated by fabricating a pressure sensor using bulk micromachining of 6H-SiC [173]. Similar to other wet etching techniques, poor directionality control is a disadvantage. Furthermore, uneven etching is observed due to shadowing of UV light by evolving microstructures as the etch progresses. The prerequisite of having electrical contact to the regions being etched limits the usefulness of electrochemical etching and PEC for surface micromachining technology because some part of the microstructure can lose electrical contact as the etching process progresses.

2.5.3 Molding

The micromolding process for SiC was developed to obtain thicker 3D microstructures of SiC. This method was very attractive during the initial development phase of SiC MEMS because no viable DRIE process was available for SiC. The micromolding process resembles macro scale SiC molding using an investment cast technique. It starts with micromold fabrication, generally using silicon DRIE. The mold is subsequently filled by depositing SiC, typically using CVD, followed by chemical-mechanical polishing to remove excess material. Finally, the mold is removed by dissolving the mold materials to release the microstructure. The process steps are schematically depicted in Figure 2.36.

The first reported SiC microstructure created using the micromolding process was for the fabrication of SiC fuel atomizers for gas turbine engines [174]. In this case, a Si mold was fabricated using a Si DRIE process and SiC was deposited using APCVD. Excess SiC was removed by mechanical lapping before releasing the structure by dissolving the Si mold with heated KOH. Figure 2.37 shows the Si mold

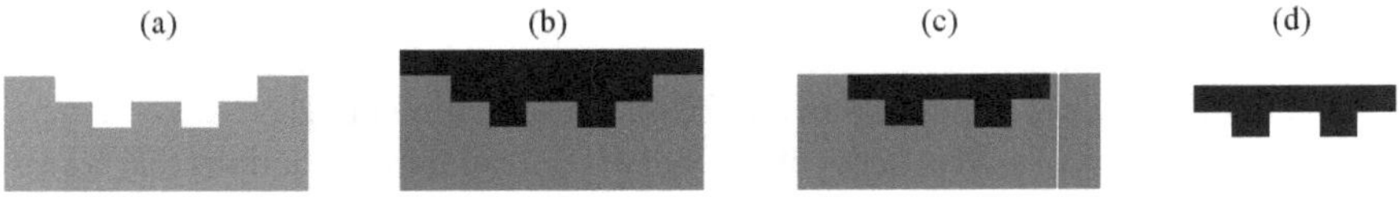

Fig. 2.36 [174] (a) micromold, (b) SiC deposited micromold, (c) micromold with SiC after removing excess SiC, and (d) released SiC microstructure (©Elsevier 1999), reprinted with permission.

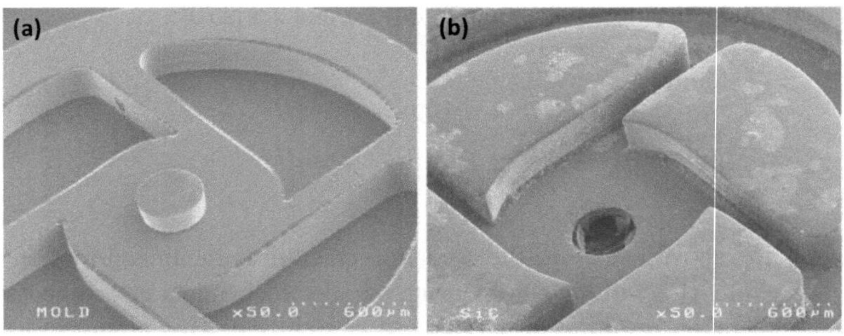

Fig. 2.37 [174] SEM image of (a) Si mold before deposition of SiC and (b) remolded SiC fuel atomizer using the Si mold (©Elsevier 1999), reprinted with permission.

and SiC fuel atomizer fabricated using the micromolding process. The performance of the micromolded SiC atomizer was compared with a Ni atomizer at similar conditions. Both atomizers performed equally well with the SiC device exhibiting a higher erosion resistance than the Ni atomizer. Some other examples of using micromolding for SiC microstructures include micromotors and micro-turbine engine parts [175, 176].

Today SiC DRIE is reasonably developed and can be successfully used to create a wide array of high aspect ratio SiC microstructures, yet there are areas of SiC microtechnology that still can benefit from the micromolding process. For instance, the maximum thickness of a bulk-micromachined SiC microstructure is predetermined by the wafer thickness. But in the molding process, thickness is determined by the mold. Furthermore, the ability to use well-established Si microfabrication techniques and no SiC etching requirement ease the fabrication constraints.

2.5.4 Other Patterning Methods for SiC

Direct writing methods such as focused ion beam (FIB) and laser micromachining have found applications in SiC microtechnology. Bhave *et al.*, for example, reports the use of FIB for creating microstructures with 195 nm electrostatic gaps for electromechanical transduction in a poly-SiC Lamé-mode resonator [127]. In terms of achievable feature size and maskless patterning ability, FIB is very desirable as a prototyping tool to validate design concepts at a low cost. The major drawback to this method is the serial nature of the process, which limits its usefulness for commercial SiC microsystem fabrication.

Laser micromachining is already widely used for cutting and drilling holes in many materials including SiC. Faster removal rate of laser micromachining is very attractive in comparison to FIB. One application is creating SiC vias of 10 to 20 μm in diameter using a femtosecond laser to bore through a 400 micron thick SiC substrate [177]. However, the holes seem to be significantly tapered and visible melting and resolidification can be observed. In theory, the use of nano- and pico-second pulse lasers will mitigate these problems, however, to date it has not yet been demonstrated for SiC patterning. Therefore, laser patterning is seen as still too immature of a process for SiC patterning to displace reactive ion etching for all but very specialized applications.

2.6 Planarization and Surface Preparation

Planarization and surface preparation of SiC are required at various stages of SiC microsystem fabrication. With regards to SiC electronics, defect-free, atomically-clean surfaces are a prerequisite for obtaining high-quality SiC epilayers because the surface defects in the substrate are replicated into the epilayer. Typically, a variety of sequential planarization and surface preparation steps are involved in SiC electronics fabrication. Although the criticality of defects and surface cleanness for MEMS is below IC standards, planarization and surface preparation steps are helpful for MEMS fabrication as well. Planarization is needed for high fidelity pattern transfer when multi-step fabrication is used. Planarization is also a necessity for the previously discussed SiC molding process. From the perspective of integrated microsystem fabrication, planarization and surface preparation will play a critical role, regardless of whether the integration scheme involves MEMS first IC second, MEMS above IC, or MEMS next to IC (see chapter 6).

Steps for atomically-cleaned planar substrate surface preparation involve lapping and mechanical polishing followed by chemical mechanical polishing and etching. Each of these steps is briefly detailed below.

2.6.1 Lapping and Mechanical Polishing

Lapping and mechanical polishing is the first step performed on as-cut wafers from the SiC boule. This step is aimed at producing wafers with minimum bow, warp, and thickness variation. Diamond or boron carbide is used as polishing material because their hardness exceeds the hardness of SiC. The mechanism of material removal from SiC by finer grit, harder abrasives occurs though mechanical microfracture because of the higher hardness of the abrasive and inherent hardness of the work material (the material that is being polished) [178]. Polishing with harder abrasive is very effective for producing wafers with minimum bow, warp, and thickness variation. It should be that polishing with hard abrasives inherently leaves scratches and residual subsurface damage that is detrimental to the subsequent SiC epitaxial process. By using abrasive slurries with decreasing grit size together with optimized rotation speed, pressure, temperature, and slurry feed, it is possible to improve the surface finish. Nonetheless, it is impossible to achieve scratch- and subsurface-defect-free results solely by using abrasive polishing [179].

2.6.2 Chemomechanical Polishing

Chemomechanical Polishing (CMP) is a two step process that starts with chemical surface modification and finishes with mechanical removal of the modified surface. Surface modification generally leads to formation of a softer passivation layer, which is subsequently removed by a softer abrasive material. As mechanical abrasion proceeds, the unreacted surface is exposed to the chemistry of the slurry and allows the chemical modification and abrasion process to repeat. As softer abrasive slurry only attacks the chemically modified surface, further scratching of hard work material, in this case SiC, is prevented.

A few different CMP processes have been reported for SiC. Generally, CMP occurs through formation of a SiO_2 passivation layer. The first CMP process, reported by Kikuchi *et al.*, uses Cr_2O_3 particles. The particles are responsible for both oxidizing the surface and mechanical abrasion [180]. A widely used SiC CMP process is based on a slurry of colloidal silica in high alkaline solution ($pH > 10$) at elevated temperatures ($\sim$55 °C) [181, 179, 101]. Availability of polishing materials and similarity to silicon processes made this combination very attractive. A polishing rate of 0.1-0.2 μmh^{-1} [181] and minimum roughness of 0.5 nm [179] were reported for this process.

Electro-chemical-mechanical means have also been exploited for improving polishing characteristics of SiC with silica slurry. A small current (1-20mA/cm^2) passes though the substrate while H_2O_2 and KNO_3 are used as the electrolytes for anodic oxidation of the SiC surface. The reported best surface roughness using this method is 0.27 nm. Although passing current through the wafer adds another parameter for processing, this method is worth consideration when superior planarity is desired.

Although further improvements in SiC CMP are desired, the current stage of SiC CMP is mature enough for surface preparation of SiC electronics and MEMS. Many industrial SiC crystal and epilayer manufacturers routinely produce defect-free SiC surfaces using CMP for epilayer growth. The current CMP processes provide needed surface finish for high quality epitaxial growth [183, 184].

2.6.3 Surface Preparation

In-situ etching with gaseous precursors is the last step of the surface preparation (see Figure 2.14) for SiC epitaxial film growth. This etching step is highly critical for epitaxial layer growth as it further reduces defects in epilayer by reducing unwanted defect-induced nucleation sites [78]. A recent report by Horita *et al.* shows clear step-and-terrace structures after the etch process that were marginally visible after the CMP process [187]. However, it is important to note that etching neither reduces the surface roughness nor scratches since etching is mostly uniform throughout the surface [188]. Therefore, the initial surface quality coming from the CMP step is critically important. Many etch chemistries have been investigated and H_2, HCl, or H_2+HCl are the most widely used etching sources, mainly due to superior etch characteristics as well as hydrogen is a common carrier gas for epilayer deposition [20, 187].

References

1. Neudeck PG (2006). Silicon Carbide Technology. The VLSI Handbook, Chapter 5 (Editor Wai-Kai Chen, CRC Press, Second Edition).
2. Rao MV, Tucker JB, Ridgway MC, Holland OW, Capano OW, Papanicolaou N, and Mittereder J (1999). Ion-implantation in bulk semi-insulating 4H-SiC. Journal of Applied Physics 86(2):752–758.
3. Pirouz P, Chorey CM, Powell JA (1987). Antiphase Boundaries in Epitaxially Grown Beta-SiC. Applied Physics Letters 50(4):221–223
4. Acheson AG (1892) British Patent 17:91
5. Tsvetkov VF, Allen ST, Kong HS, Carter Jr CH (1996). Institute of Physics Conference Series 142: 17–22
6. Tairov YM, Tsvetkov VF (1983). Progress in controlling the growth of polytypic crystals.Progress In Crystal Growth And Characterization 7(1-4):111–162
7. Sudarshan TS, Maximenko SI (2006). Bulk growth of single crystal silicon carbide. Microelectronic Engineering 83:155159
8. Lely JA (1955). Darstellung von Einkristallen von Siliziumcarbid und Beherrschung von Art und Menge der eingbauten Verunreinigungen. Berichte Deutche Keramik Geselshaft (32):229–231
9. Tairov YM, Tsvetkov VF (1978). Investigations and growth processes of ingots of silicon carbide single crystals. Journal of Crystal Growth 43 (2):209–212
10. Davis RF, Carter Jr CH, Hunter CE (1995). US Patent No. Re 34861
11. Stein RA, Lanig P (1992). Influence of surface-energy on the growth of 6H-SiC and 4H-SiC polytypes by sublimation. Materials Science and Engineering B 11(1-4): 69–71.
12. Barrett DL, McHugh JP, Hobgood HM, Hopkins RH, McMullin PG, Clarke RC (1993). Growth of large SiC single crystals. Journal of Crystal Growth 128(1):358
13. Cree Inc. USA. http://www.cree.com
14. Glass RC, Henshall G, Tsvetkov VF, Carter Jr CH (1997). SiC Seeded Crystal Growth. Physica Status Solidi B 202:149-162
15. Chaussende D, Wellmann PJ, Pons M (2007). Status of SiC bulk growth processes. JOURNAL OF PHYSICS D: APPLIED PHYSICS 40(20):61506158
16. Norstel AB, Sweden. www.norstel.com

17. Chaussende D, Baillet F, Charpentier L, Pernot E, Pons M, Madara R (2003). Continuous Feed Physical Vapor Transport Toward High Purity and Long Boule Growth of SiC. Journal of The Electrochemical Society 150(10):G653-G657
18. Fanton M, Skowronski M, Snyder D, Chung H.J, Nigam S, Weiland B, Huh SW (2004). Growth of Bulk SiC by Halide Chemical Vapor Deposition. Materials Science Forum 457-460:87–90
19. Wellmann P, Desperrier P, Muller R, Straubinger T, Winnacker A, Baillet F, Blanquet E, Dedulleb JM, Pons M (2005).SiC single crystal growth by a modified physical vapor transport technique. Journal of Crystal Growth 275(1-2):e555e560
20. ltoh A, Matsunami H (1997). Single crystal growth of SiC and electronic devices. Critical Reviews in Solid State and Materials Sciences, 22(2):111–197
21. Takahashi J, Ohtani N, Kanaya M (1995). Influence of the Seed Face Polarity on the Sublimation Growth of alpha-SiC. Japanese Journal of Applied Physics 34(9A):4694–4698
22. Ito A, Kimoto T, Matsunami H (1994). High-quality 4H-SiC homoepitaxial layers grown by step-controlled epitaxy. Applied Physics Letters 65(11):1400–1402
23. Kordina O, Hallin C, Ellison A, Bakin AS, Ivanov IG, Henry A, Yakimova R, Tuominen M, Vehanen A, Janzen E (1996): Applied Physice Letters 69(10):1456–1458
24. Ellison A, Magnusson B,Sundqvist B, Pozina G, Bergman JP, Janzen E, Vehanen A (2004). SiC crystal growth by HTCVD. Materials Science Forum 457-460:9–14
25. Ellison A, Magnusson B, Son NT, Storasta L, Janzen E (2003). HTCVD grown semi-insulating SiC substrates. Materials Science Forum 433-436:33-38
26. Sundqvist B, Ellison A, Jonsson A,Henry A,Magnusson B, Janzen E (2003). Growth of High Quality p-type 4H-SiC Substrates by HTCVD. Materials Science Forum 433-436:21–24
27. Schmid F and Pensl G (2004). Comparison of the electrical activation of P+ and N+ ions co-implanted along with Si+ or C+ ions into 4H-SiC. Applied Physics Letters 84(16):3064–3066
28. Chaussende D, Ucar M, Auvray L, Baillet F, Pons M, Madar R (2005). Control of the Supersaturation in the CF-PVT Process for the Growth of Silicon Carbide Crystals: Research and Applications. Crystal Growth & Design 5(4):15391544.
29. Chaussende D, Eid J, Mercier F, Madar R, Pons M (2009). Nucleation and Growth of 3C-SiC Single Crystals from the Vapor Phase. Materials Science Forum 615-617:31–36
30. Polyakov AY, Fanton MA, Skowronski M, Chung HJ, Nigaml S, Huh SW (2006). Halide-CVD Growth of Bulk SiC Crystals. Materials Science Forum 527-529:21–26
31. Hofmann D, Schmitt E, Bickermann M, Kolbl M, Wellmann PJ, Winnacker A (1999). Prospects of the use of liquid phase techniques for the growth of bulk silicon carbide crystals. Materials Science and Engineering B6162:2939
32. Halden F, Meuli WP, Fredericks WJ (1961), Growth of SiC Single Crystals. Final Technical Report. contract. no. NObsr-72772, Bureau of Ships.
33. Ujihara T, Munetoh S, Kusunoki K, Kamei K, Usami N, Fujiwara K, Sazakia G, Nakajima K (2005). Crystal quality of a 6H-SiC layer grown over macrodefects byliquid-phase epitaxy: a Raman spectroscopic study. Thin Solid Films 476(1):206209
34. Kusunoki K, Kamei K, Ueda Y, Naga S, Ito Y, Hasebe M, Ujihar T Nakajima K (2005). Crystalline Quality Evaluation of 6H-SiC Bulk Crystals Grown from Si-Ti-C Ternary Solution. Materials Science Forum 483-485:13-16
35. Eid J, Santailler JL, Ferrand B, Basset A, Passero A, Lewandowska R, Balloud C, Camassel J (2006). Improvement of cubic silicon carbide crystals grown from solution. Superlattices and Microstructures 40:201204
36. Kamei K, Kusunoki K,Yashiro N, Okada N, Tanaka T,Yauchi A (2009). Solution growthofsinglecrystalline6H,4H-SiCusingSiTiCmelt. Journal of Crystal Growth 311(3):855858
37. Ujihara T, Maekawa R, Tanaka R, Sasaki K, Kuroda K, Y. Takeda (2008). Solution growth of high-quality 3C-SiC crystals. Journal of Crystal Growth 310 (7-9):14381442
38. Ohtani N, Katsuno M, Tsuge H, Fujimoto T, Nakabayashi M, Yashiro H, Sawamura M, Aigo T, Hoshino T (2006). Dislocation processes during SiC bulk crystal growth. Microelectronic Engineering 83(1):142145

39. Rost HJ, Doerschel J, Schulz D, Siche D, Wollweber J (2002). Microdefect Generation in Single Crystal SiC Caused by Polytype Changes. Materials Science Forum 389-393:67-70
40. Hofmann D, Schmitt E, Bickermann M, Kolbl M, Wellmann PJ, Winnacker A (1999). Analysis on defect generation during the SiC bulk growth process. Materials Science and Engineering B6162:48–53
41. Yakimova R, Iakimov T, Syvajarvi M, Jacobsson H, Raback P, Vehanen A, Janzen E (1999). Polytype stability and defect reduction in 4H-SiC crystals grown via sublimation technique. MRS Symposium Proceedings 572:265-267
42. Sanchez E, Kopec A, Poplawski S, Ware R, Holmes S, Wang S, Timmerman A (2002). The Nucleation of Polytype Inclusions during the Sublimation Growth of 6H and 4H Silicon Carbide. Materials Science Forum 389393:71-74
43. Kanaya M, Takahashi J, Fujiwara Y, Moritani A (1991). Controlled sublimation growth of single crystalline 4H-SIC and GH-SiC and identification of polytypes by x-ray diffraction. Applied Physics Letter 58(1):56-58
44. Yakimova R, Syvajarvi M, Iakimov T, Jacobsson H, Raback P, Vehanen A, Janzen E (2000). Polytype stability in seeded sublimation growth of 4H-SiC boules.Journal of Crystal Growth 217:255-262
45. Tupitsyn EY, Arulchakkaravarthi A, Drachev RV, Sudarshan TS (2007). Controllable 6H-SiC to 4H-SiC polytype transformation during PVT growth. Journal of Crystal Growth 299(1):7076.
46. Rost HJ, Doerschel J, Irmscher K, Rossberg M, Shulz D, Siche D (2005). Polytype stability in nitrogen-doped PVT grown 2 inch 4HSiC crystals. Journal of Crystal Growth 275(1-2):e451-e454
47. Vodakov YA, Mokhov EN, Roenkov AD, Anikin MM (1979). Effect of impurities on the polymorphism of silicon carbide. Soviet Technical Physics Letters 5(3):147-148
48. Rost HJ, Schmidbauer M, Siche D, Fornari R (2006). Polarity- and orientation-related defect distribution in 4H-SiC single crystals. Journal of Crystal Growth 290(1):137-143
49. Siche D, Albrecht M, Doerschel J, Irmscher K, Rost HJ, Roxberg M, Schulz D (2004). Effect of Nitrogen Doping on the Formation of Planar Defects in 4H-SiC. Materials Science Forum 483485:39-42
50. Ohtani N, Takahashi J, Katsuno M, Yashiro H, Kanaya M (1998). Development of Large Single-Crystal SiC Substrates. Electronics and Communications in Japan Part 2 81(6):8-19
51. Fissel A (2000). Thermodynamic considerations of the epitaxial growth of SiC polytypes. Journal of Crystal Growth 212 (3-4):438-450
52. Schulze N, Barrett D, Weidner M, Pensl G (2000). Controlled Growth of Bulk 15R-SiC Single Crystals by the Modified Lely Method. Materials Science Forum 338-342:111-114
53. Neudeck PG, Powell JA (1994). Performance limiting micropipe defects in silicon carbide wafers. IEEE Electron Device Letters 15(2):63-65
54. Frank FC (1951). Capillary equilibria of dislocated crystals. Acta Crystallographica 4(6):497-501
55. Dudley M, Huang XR, Huang W, Powell JA, Wang S, Neudeck P, Skowronski M (1999). The mechanism of micropipe nucleation at inclusions in silicon carbide. Applied Physics Letters 75(6):784-86
56. Tsvetkov VF, Allen ST, Kong HS, Carter Jr CH (1996). Recent progress in SiC crystal growth. Institute of Physics Conference Series (142):17-22
57. Schulze N, Barrett DL, Pensl G (1998). Near-equilibrium growth of micropipe-free 6H-SiC single crystals by physical vapor transport. Applied Physics Letters 72(13):1632-1634
58. Takahashi J, Ohtani N, Kanaya M (1996). Structural defects in alpha-SiC single crystals grown by the modified-Lely method. Journal of Crystal Growth 167(3-2):596-606
59. Takahashi J, Kanaya M, Hoshino T (1994). Sublimation growth and characterization of SiC single crystalline ingots on faces perpendicular to the (0001) basal plane. Institute of Physics Conference Series 137:13-16
60. Takahashi J, Ohtani N (1997). Modified-Lely SiC Crystals Grown in [1100] and [1120] Directions. Physica Status Solidi B 202(1):163-175

61. Takahashi J, Ohtani N, Katsuno M, Shinoyama S (1997). Sublimation growth of 6H- and 4H-SiC single crystals in the [1100] and [1120] directions. Journal of Crystal Growth 18(3):229-240
62. Nakamura D, Gunjishima I, Yamaguchi S, Ito T, Okamoto A, Kondo H, Onda S, Takatori K (2004). Ultrahigh-quality silicon carbide single crystals. Nature 430:1009-1012
63. Carter, Jr CH, Tsvetkov VF, Glass RC, Henshall D, Brady M, Muller StG, Kordina O, Irvine K, Edmond JA, Kong HS, Singh R, Allen ST, Palmour JW (1999). Progress in SiC: from material growth to commercial device development. Materials Science and Engineering B6162:18
64. Leonard RT, Khlebnikov Y, Powell AR, Basceri C, Brady MF, Khlebnikov I, Jenny JR, Malta DP, Paisley MJ, Tsvetkov VF, Zilli R, Deyneka E, Hobgood HMcD, Balakrishna V, Carter Jr CH (2009). 100 mm 4HN-SiC Wafers with Zero Micropipe Density. Materials Science Forum 600-603:7-10
65. Kong HS , Glass JT, Davis RF (1989). Growth rate, surface morphology, and defect microstructures of SiC films chemically vapor deposited on 6HSiC substrates. Journal of Materials Research 4(1):204214
66. Zorman CA, Fleischman AJ, Dewa AS, Mehregany M, Jacob C, Nishino S and Pirouz P (1995). Epitaxial growth of 3C-Sic films on 4 in. diam (100) silicon wafers by atmospheric pressure chemical vapor deposition. Journal of Applied Physics 78(8):5136-5138
67. Pazik JC, Kelner G, Bottka N (1991). Epitaxial growth of SiC on silicon-on-sapphire substrates by chemical vapor deposition. Applied Physics Letters 58(13):14191422
68. Givargizov EI (1975). Fundamental Aspects of VLS Growth. Journal of Crystal Growth 31:20-30
69. Milewski JV, Gac FD, Petrovic SR, Skaggs SR(1985). Growth of beta-SiC Whiskers by the VLS process. Journal of materials Science 20:1160-1166
70. Kong HS, Glass JT, Davis RF (1991). US Patent No. 5011549
71. Rupp R, Wiedenhofer A, Stephani D (1999). Epitaxial growth of SiC in a single and a multi wafer vertical CVD system: a comparison. Materials Science and Engineering B61-62:125-129
72. Karlsson S, Nordell N, Spadafora F, Linnarsson M (1999). Epitaxial growth of SiC in a new multi-wafer VPE reactor. Materials Science and Engineering B61-62:143146
73. Thomas B, Bartsch W, Stein R, Schrner R, and Stephani D (2004). Properties and Suitability of 4H-SiC Epitaxial Layers Grown at Different CVD Systems for High Voltage Applications. Materials Science Forum 457-460:181-184
74. Powell JA, Larkin DJ (1997). Process-Induced Morphological Defects in Epitaxial CVD Silicon Carbide. Physica Status Solidi B 202 (1):529-548
75. Via F L, Izzo G, Abbondanza G, Crippa D (2009). Thick Epitaxial Layers Growth by Chlorine Addition. Materials Science Forum 615-617:55-60
76. Ellison A, Zhang J, Henry A, Janzen E (2002). Epitaxial growth of SiC in a chimney CVD reactor. Journal of Crystal Growth 236(1-3):225238
77. Rupp R, Makarov YN, Behner H, Wiedenhofer A (1997). Silicon Carbide Epitaxy in a Vertical CVD Reactor: Experimental Results and Numerical Process Simulation. Physica Status Solidi B 202 (1):281-304
78. Burk AA (2006). Development of Multiwafer Warm-Wall Planetary VPE Reactors for SiC Device Production. Chemical Vapour Deposition 12:465473
79. Powell JA, Rowland LB (2002). SiC Materials-Progress, Status, and Potential Roadblocks Proceedings of the IEEE 90(6):942-955
80. O. Kordina, Hallin C, Glass RC, Janzen E (1994). Proceedings of the International Conference on SiC, Inst. Phys. Conf. Ser.137: 41
81. Kordina O, Hallin C, Henry A, Bergman JP, Ivanov IG, A, Ellison A, Son NT, Janzen E (1997). Growth of SiC by Hot-Wall CVD and HTCVD. Physica Status Solidi B 202(1): 321-334
82. Frijlink PM (1988). A new versatile, large size MOVPE reactor. Journal of Crystal Growth 93(1-4):207-215

83. Burk AA, OĹoughlin MJ, Nordby Jr HD (1999). SiC epitaxial layer growth in a novel multiwafer vapor-phase epitaxial (VPE) reactor Journal of Crystal Growth 200:458-466
84. Burk AA, OĹoughlin MJ, Sumakeris JJ, Hallin C, Berkman E, Balakrishna V, Young J, Garrett L, Irvine KG, Powell AR, Khlebnikov Y, Leonard RT, Basceri C, Hull BA, Agarwal AK (2009). SiC Epitaxial Growth on Multiple 100-mm Wafers and its Application to Power-Switching Devices. Materials Science Forum Vols. 600-603:77-82
85. Ito M, Storasta L, Tsuchida H (2008). Development of 4H-SiC Epitaxial Growth Technique Acheiving High Growth Rate and Large-Area Uniformity. Applied Physics Express 1:015001-1
86. Shibahara K, Nishino S, Matsunami H (1987). Antiphase domain free growth of cubic SiC on Si (100). Applied Physics Letters 50 (26):1888–1890
87. Kong HS, Glass JT, Davis RF (1988). Chemical vapor deposition and characterization of 6HSiC thin films on off-axis 6HSiC substrates. Journal of applied physics 64(5): 2672–2679
88. Powell J A, Larkin DJ, Matus, LG, Choyke WJ, Bradshaw, JL, Henderson L, Yoganathan M, Yang J, Pirouz P (1990). Growth of high-quality 6HSiC epitaxial films on vicinal (0001) 6HSiC wafers. Applied Physics Letters 56(15):14421444
89. Kimoto T, Matsunami H (1994). Surface kinetics of adatoms in vapor phase epitaxial growth of SiC on 6H-SiC(0001) vicinal surfaces. Journal of Applied Physics 75 (2):850–859
90. Kimoto T, Nishino H, Yoo WS, Matsunami H, Nishino H (1993). Growth mechanism of 6H-SiC in step-controlled epitaxy. Journal of Applied Physics 73(2):726–732
91. Larkin DJ, Neudeck PG, Powell AJ, Matus LG (1994). Site-competition epitaxy for superior silicon carbide electronics. Applied Physics Letters 65(13):1659–1661
92. Choyke WJ (1990). The Physics and Chemistry of Carbides, Nitrides, and Borides, NATO AS1 Series E: Applied Sciences, edited by R. Freer (Khrwer.Dordrecht. 1990). 185:853
93. Davis RF, Glass JT (1991). Advances in Solid State Chemistry, edited by Catlow CRA (JAI, Greenwich, CT). 2:l–111
94. Larkin DJ (1997). SiC Dopant Incorporation Control Using Site-Competition CVD. Physica Status Solidi B (b) 202 (1):305–329
95. Syvajarvi M, Yakimova R, Tuominen M, Kakanakova-Georgieva A, MacMillan MF, Henry A, Wahab Q, Janzen E (1999). Growth of 6H and 4H-SiC by sublimation epitaxy. Journal of Crystal Growth 197 (1): 155–162
96. Dmitriev V (1995), LPE of SiC and SiC-AlN, in Properties of Silicon Carbide. ser. 13, G. L. Harris, Ed. London, U.K.: INSPEC
97. Syvajarvi M, Yakimova R, Radamson HH, Son NT, Wahab Q, Ivanov IG, Janzen E (1999). Liquid phase epitaxial growth of SiC. Journal of Crystal Growth 197 (1):147–154
98. Nishitani SR, Kaneko T (2008). Metastable solvent epitaxy of SiC. Journal of Crystal Growth 310(7-9):1815–1818
99. Ferro G, Jacquier C (2004). Groth by a vapour-liquid-solid mechanism: a new approach for silicon carbide epitaxy. New Journal of Chemistry 28:889–896
100. Dowcorning.com
101. Nipponsteel.com
102. Wu CH, Zorman CA, Mehregany M (2000). Characterization of polycrystalline SiC grown on SiO2 and Si3N4 by APCVD for MEMS applications. Materials Science Forum 338-342:541–544
103. Cheng L, Pan M, Scofield J, Steckl AJ (2002). Growth and Doping of SiC-Thin Films on Low-Stress, Amorphous Si3N4/Si Substrates for Robust Microelectromechanical Systems Applications. Journal of Electronic Materials 31(5):361-365
104. Stoldt CR, Carraro C, Ashurst WR, Gao D, Howe RT, Maboudian R (2002). Low temperature CVD process for SiC MEMS. Sensors and Actuators A 97-98:410-415
105. Ashurst WR, Wijesundara MBJ, Carraro C, Maboudian R (2004). Tribological Impact of SiC Encapsulation of Released Polycrystalline Silicon Microstructures. Tribology Letters 17:195-198
106. Song X, Rajgopal S, Melzak J, Zorman CA, M. Mehregany M (2002). Development of a multilayer SiC surface micromachining process with capabilities and design rules comparable to conventional polysilicon surface micromachining. Materials Science Forum 389-393:755-758

107. Zorman CA, Fleischman AJ, Dewa AS, Mehregany M, Jacob C, Nishino S, Pirouz P (1995). Epitaxial growth of 3C-SiC films on 4 in. diam (100) silicon wafers by atmospheric pressure chemical vapor deposition. Journal of Applied Physics 78 (8):5136-5138
108. Fleischman AJ, Roy S, Zorman CA, Mehregrany (1996). Polycrystalline silicon carbide for surface micromachining. Proceedings of 9th Annual. International Workshop on Microelectromechanical Systems, San Diego, CA, Feb. 1115 1996 :473478
109. Chung GS, Kim KS, Han KB (2008). Characteristics of polycrystalline 3C-SiC thin films grown on Si wafers for harsh environment microdevices. Ceramics International 34:841844
110. Kim KS, Chung GS (2009). Growth and characteristics of polycrystalline 3CSiC films for extreme environment micro/nano-electromechanical systems. Sensors and Actuators A 155:125130
111. Zhang J, Howe RT, Maboudian R (2006). Control of strain gradient in doped polycrystalline silicon carbide films through tailored doping. J. Micromech. Microeng. 16:L1-L5
112. Lee KW, YU KS, Kim Y (1997). Heretoepitaxial growth of 3C-SiC on Si(001) without carbonization. Journal of Crystal Growth 179(1-2):153-160
113. Hurtos E, Rodriguez-Viejo J (2000). Residual stress and texture in poly-SiC films grown by low-pressure organometallic chemical-vapor deposition. Journal of Applied Physics 87(4):1748-1758
114. Wang CF, Tsai DS (2000). Low pressure chemical vapor deposition of silicon carbide from dichlorosilane and acetylene. Materials Chemistry and Physics 63:196-201
115. Stoldt CR, Fritz MC, Carraro C, Maboudian R (2001). Micromechanical properties of silicon-carbide thin films deposited using single-source chemical-vapor deposition. Applied Physics Letters 79(3):437-349
116. Wijesundara MBJ, Valente G, Ashurst WR, Howe RT, Pisano AP, Carraro C, Maboudian R (2004). Single-Source Chemical Vapor Deposition of 3C-SiC Films in a LPCVD Reactor Part I: Growth, Structure, and Chemical Characterization. Journal of the Electrochemical Society 151:C210-C214
117. Valente G, Wijesundara MBJ, Carraro C, Maboudian R (2004). Single-Source Chemical Vapor Deposition of 3C-SiC Films in a LPCVD Reactor Part II: Reactor Modeling and Chemical Kinetics. Journal of the Electrochemical Society 151:C 215-C219
118. Zorman CA, Rajgopal S, Fu XA, Jezeski R, Melzak J, Mehregany M (2002). Deposition of Polycrystalline 3C-SiC Films on 100 mm Diameter Si.100. Wafers in a Large-Volume LPCVD Furnace. Electrochemical and Solid-State Letters 5(10):G99-G101
119. Wijesundara MBJ, Stoldt CR, Carraro C. Howe RT, Maboudian R (2002). Nitrogen Doping of Polycrystalline 3C-SiC Films Grown by Single-Source Chemical Vapor Deposition. Thin Solid Films 419:69-75
120. Wijesundara MBJ, Gao D, Carraro C. Howe RT, Maboudian R (2003). Nitrogen Doping of Polycrystalline SiC Films Grown using 1,3Disilabutane in conventional LPCVD Reactor. Journal of Crystal Growth 259:18-25
121. Gao D, Wijesundara MBJ, Howe RT, Maboudian R (2003). Characterization of residual strain in SiC films deposited using 1,3-disilabutane for MEMS Application. Journal of Microlithography Microfabrication and Microsystems 2:259-264
122. Roper CS, Carraro C. Howe RT, Maboudian R (2006). Silicon carbide thin films using 1,3-disilabutanesingle precursor for MEMS ApplicationsReview. ESC Transctions 3(10):267-280
123. Zhang J, Howe RT, Maboudian R (2006). Electrical Characterization of n-Type Polycrystalline 3C-Silicon Carbide Thin Films Deposited by 1,3-Disilabutane. Journal of The Electrochemical Society, 153(6):G548-G551
124. Wijesundara MBJ, Walther DC, Stoldt CR, Fu K, Gao D, Carraro C, Pisano AP, Maboudian R (2003). Low Temperature CVD SiC Coated Si Microcomponents for Reduced Scale Engines, ASME International Mechanical Engineering Congress 2003 2:41696
125. Azevedo RG, Zhang J, Jones DG, Myers DR, Jog AV, Jamshidi B, Wijesundara MBJ, Maboudian R, Pisano AP(2007). Silicon Carbide Coated MEMS Strain Sensor for Harsh Environment Applications, MEMS 2007, 20th IEEE International Conference on Micro Electro Mechanical Systems, Technical Digest 2007:643-646

126. Jamshidi B, Azevedo RG, Wijesundara MBJ, Pisano AP (2007). Corrosion Enhanced Capacitive Strain Gauge at 370C. MEMS 2007, 20th IEEE SENSORS 2007 Conference on Micro and Nano sensors Technical Digest 2007:804-807
127. Bhave SA, Gao D, Maboudian R, Howe RT (2005). Fully-differentical poly-SiC lame-mode resonator and checkerboard filter. MEMS 2005, 18th IEEE International Conference on Micro Electro Mechanical Systems, Technical Digest (2005):223-226
128. Liu F, Carraro C, Chu J, Maboudian R (2009). Residual stress characterization of polycrystalline 3C-SiC films on (Si 100) deposited from methylsilane. Journal of Applied Physics 106:013505
129. Liu F, Carraro C, Chu J, Pisano AP, Maboudian R (2010). Growth and characterization of nitrogen-doped polycrystalline 3C-SiC thin films for harsh environment MEMS applications. Journal of Micromechanics Microengineering 20:035011
130. Fu XA, Trevino J, Mehregany M (2006). Nitrogen-doping of polycrystalline 3C-SiC films deposited by low pressure chemical vapor deposition. Materials Science Forum 527-529:311-314
131. Myers DR, Cheng KB, Jamshidi B, Azevedo RG, Senesky DG, Wijesundara MBJ Pisano AP (2009). A Silicon Carbide Resonant Tuning Fork for Micro-Sensing Applications in High Temperature and High G-Shock Environment. Journal of Micro/Nanolithography, MEMS, and MOEMS 8:021116
132. Rajgopal S, Zula D, Garverick S, Mehregany M (2009). A Silicon Carbide Accelerometer for Extreme Environment Applications. Materials Science Forum 600-603:859–862
133. Chen L, Mehregany M (2008). A silicon carbide capacitive pressure sensor for in-cylinder pressure measurement. Sensors and Actuators A 145146:2-8
134. Chen J, Steckel AJ, Loboda MJ (2000). In situ N-2-doping of SiC films grown on Si(111) by chemical vapor deposition from organosilanes. Journal of Electrochemical Society 147:2324–2327
135. Kern RS, Davis RF (1997). Deposition and doping of silicon carbide by gas-source molecular beam epitaxy Appl. Phys. Lett. 71(10):1356-1358
136. Chang WT. Zorman C (2009). Grain size control of (111) polycrystalline 3C-SiC films by doping used as folded-beam MEMS resonators for energy dissipation. Microsystem Technology 15:875–880
137. Murooka KI, Higashikawa I, Gomei Y (1996). Improvement of the Youngs modulus of SiC film by low-pressure chemical vapor deposition with B2H6 gas. Applied Physics Letters 69(1):37–39
138. Fu XA, Dunning J, Zorman CA, Mehregany M (2004). Development of a High-Throughput LPCVD Process for Depositing Low Stress Poly-SiC. Materials Science Forum 457-460:305–308
139. Roper CS, Radmilovic V, Howe RT, Maboudian R (2008). Characterization of polycrystalline 3C-SiC films deposited from the precursors 1,3-disilabutane and dichlorosilane. Journal of Applied Physics 103:084907
140. Fu XA, Dunning J, Zorman CA, Mehregany M (2005). Polycrystalline 3C-SiC thin films deposited by dual precursor LPCVD for MEMS applications. Sensors and Actuators A 119:169-176
141. Zhang J, Ph.D. Thesis (UC Berkeley)
142. Azevedo RG, Jones DG, Jog AV, Jamshidi B, Myers DR, Chen L, Fu, XA, Mehregany, M, Wijesundara, MBJ, Pisano, AP (2007). A SiC MEMS Resonant Strain Sensor for Harsh Environment Applications, IEEE Sensors, 7(4):568–576
143. Sarro PM (2000). Silicon carbide as a new MEMS technology. Sensors and Actuators 82:210218
144. Sarro PM, deBoer CF, Korkmaz E, Laros JMW (1998). Low-stress PECVD SiC thin films for IC-compatible microstructures. Sensors and Actuators A 67:175–180
145. Pelegrini MV, Rehder GP, Pereyra L (2010). a-SiC:H films deposited by PECVD for MEMS applications. Physica Status Solidi C 7:786–789
146. Shimizu H, Kato A (2009). Low temperature growth of 3C-SiC Films on (111) by Plasma Assisted CVD. Materials Science Forum 615-617:161–164

147. Rajaraman V, Pakula LS, Pham HTM, Sarro PM, French PJ (2009). RobustWafer-Level Thin-Film Encapsulation of Microstructures Using Low Stress PECVD Silicon Carbide. MEMS 2009, Sorrento, Italy, Jan. 25-29 2009:140-143
148. Gonzalez-Elipe AR, Yubero F, Sanz JM (2003). Low Energy Ion Assisted Film Growth. Imperial College Press, London, UK
149. Valentini A, Convertino A, Alvisi AM, Cingolani R, Ligonzo T, Lamendola R, Tapfer L (1998). Synthesis of silicon carbide thin films by ion beam sputtering. Thin Solid Films 335:80–84
150. Pezoldt J, Stottko B, Kupris G, Ecke G (1995). Sputtering effects in hexagonal silicon carbide. Materials Science and Engineering B29:94–98
151. Zaytouni M, Riviere JP, Denanot MF, Allain J (1996). Structural characterization of SiC films prepared by dynamic ion mixing. Thin Solid Films 287:1–7
152. Jones DG, Azevedo RG, Chan MW, Pisano AP, Wijesundara MBJ (2007). Low-Temprature Ion Beam Sputter Deposition of Amorphous Silicon Carbide for Wafer Level Vacuum Sealing, MEMS 2007, 20th IEEE International Conference on Micro Electro Mechanical Systems, Technical Digest 2007: 275–278
153. Argyrakis P, McNabb P, Snell AJ, Cheung R (2006). Relaxation of process induced surface stress in amorphous silicon carbide thin films using low energy ion bombardment. Applied Physics Letters 89:034101
154. Park WT, Candler R, Kronmueller S, Lutz M, Partridge A, Yama G, Kenny T(2003). Wafer-scale film encapsulation of micromachined accelerometers. Proc. of International Conference on Solid State Sensors, Actuators and Microsystems (Transducers 03). IEEE, 2003:
155. Rusu M, Jansen H, Gunn R, Witvrouw A (2004). Self-aligned 0-level sealing of MEMS devices by a two layer thin film reflow process. Microsystem Technologies 10:364371
156. Roper CS, Candler R,Yoneoka S, Kenny T Howe RT, Maboudian R (2009). Simultaneous Wafer-Scale Vacuum Encapsulation and Microstrcture Cladding with LPCVD 3C-SiC. Transducers 2009, Denver, CO, USA, June 21-25 2009:1031–1034
157. Fraga MA, Massi M, Oliveira IC, Maciel HS, Filho SGDS, Mansano RD (2008). Nitrogen doping of SiC thin films deposited by RF magnetron sputtering. Journal of Materials Science: Materials Electronics 19:83-5840
158. Serre C, Perez-Rodriguez A, Morante JR, Esteve J, Acero MC, Kogler R, Skorupa W (2000). Ion beam synthesis of polycrystalline SiC on SiO2 structures for MEMS applications. Journal of Micromechanics Microengineering 10:152-156
159. Yih PH, Saxena V, Steckl AJ (1997). A Review of SiC Reactive Ion Etching in Fluorinated Plasmas. Physica Status Solidi B 202(1):605–642
160. Pan WS, Steckl AJ (1990). Ion Etching of SiC Thin Films by Mixtures of Fluorinated Gases and Oxygen. Journal of the Electrochemical Society 137(1):212–220
161. Gao D, Wijesundara MBJ, Carraro C, Howe RT, Maboudian R (2004). Transformer coupled plasma etching of 3C-SiC films using fluorinatedchemistry for microelectromechanical systems applications. Journal of Vacuum Science Technology B 22(2):513–518
162. Mayer TM, Barker RA (1982). Simulation of plasma-assisted etching processes by ion-beam techniques. Journal of Vacuum Science and Technology 21(3):757–763
163. Lee HY, Kim DW, Sung YJ, Yeom GY (2005). Fabrication of SiC micro-lens by plasma etching. Thin Solid Films 475(1-2):318322
164. Gao D, Howe RT, Maboudian R (2003). High-selectivity etching of polycrystalline 3C-SiC films using HBr-based transformer coupled plasma. Applied Physics Letter 82(11):1742–1744
165. Chabert P (2001). Deep etching of silicon carbide for micromachining applications:Etch rates and etch mechanisms. Journal of Vacuum Science Technology B 19(4):1339–1345
166. Plank NOV, Blauw MA, van der Drift EWJM, Cheung R (2003). The etching of silicon carbide in inductively coupled SF6/O2 plasma. Journal of Applied Physics D: Applied Physics 36:482–487
167. SPP Process Technology Systems (SPTS), UK. http://www.spp-pts.com
168. Fleischman AJ, Zorman CA, Mehregany M (1998). Etching of 3C-SiC using CHF3/O2 and CHF3/O2/He plasmas at 1.75 Torr. Journal of Vacuum Science Technology B 16(2):536-539

169. Oxford Instruments, UK. http://www.oxford-instruments.com
170. Zhuang D, Edgar JH (2005). Wet etching of GaN, AlN, and SiC: a review. Materials Science and Engineering R 48(1):1-46
171. Alok D, Baliga BJ (1995). A Novel Method for Etching Trenches in Silicon Carbide. Journal of Electronic Materials 24:311
172. Roper SR, Howe RT, Maboudian R (2009). Room-Temperature Wet Etching of Polycrystalline and Nanocrystalline Silicon Carbide Thin Films with HF and HNO3. Journal of The Electrochemical Society 156 (3):D104-D107
173. R. Okojie, A. Ned, A. Kurtz, and W. Carr (1996). 6H-SiC pressure sensors for high temperature applications. Proceeding of 9th Annual International Workshop Microelectromechanical Systems, M. Allen and M. Reed, Eds., San Diego, CA, Feb. 1115 1996:146-149
174. Mehregany M, Zorman CA (1999). SiC MEMS: opportunities and challenges for application in harsh environments. Thin Solid Films 355-356:518–524
175. Yasseen A, Zorman CA, Mehregany M (1999). Surface Micromachining of Polycrystalline SiC Films Using Microfabricated Molds of SiO and Polysilicon. Journal of Microelectromechanical system 8 (3):237–242
176. Lohner KA, Chen KS, Ayon AA, Spearing SM (1998). Microfabricated Silicon Carbide Microengine Strucutres. Materials Research Society Symposium Proceedings Series 546:1–6
177. Farsari M, Filippidis G, Zoppe S,Reider GA, Fotakis C (2005). Efficient femtosecond laser-micromachining of bulk 3C-SiC Journal Micromechanics and Microengineering 15:1786-1789
178. Jiang M, Komanduri R (1998). On the finishing of Si3N4 balls for bearing applications. Wear 215(1-2):267–278
179. Chen X, Li J, Ma, Hu X, Xu X, Jiang M (2006). Fine Machining of Large-Diameter 6H-SiC Wafers. Journal of Materials Science and Technology 22(5):681–684
180. Kikuchi M, Takahashi Y, Suga T, Suzuki S, Bando Y(1992). Mechanochemical Polishing of Silicon Carbide Single Crystal with Chromium(III) Oxide Abrasive. Journal of American Ceramic Society 75 (1):189–194
181. Zhou L, Audurier, Pirouz P, Powell J A (1997). Chemomechanical Polishing of Silicon Carbide. Journal of the Electrochemical. Society 144(6):L161–163
182. Li C, Bhat IB, Wang R, Seiler J (2004). Electro-Chemical Mechanical Polishing of Silicon Carbide. Journal of Electronic Materials 33(5):481–486
183. NOVASiC, France. http://www.novasic.com
184. Anderson TA, Barrett DL, Chen J, Elkington WT, Emorhokpor E, Gupta A, Johnson CJ, Hopkins RH, Martin C, Kerr T, Semenas E, Souzis AE, Tanner CD, Yoganathan M, Zwieback I (2004). Advanced PVT Growth of 2 & 3-Inch Diameter 6H SiC Crystals. Materials Science Forum 457-460:75–78
185. Fu XA, Zorman CA, Mehregany M (2002). Chemical Mechanical Polishing of Cubic Silicon Carbide Films Grown on Si(100) Wafers. Journal of the Electrochemical Society 149(12):G643–G647
186. Burk AA, Rowland LB (1996). Novel in situ optical monitoring method for selective area metalorganic vapor phase epitaxy. Journal of Crystal Growth 167(3-4):586–595
187. Horita M, Kimoto T, Suda (2008). Surface Morphologies of 4H-SiC (1120) and (1100) Treated by High-Temperature Gas Etching. Japanese Journal of Applied Physics 47(11):8388–8390
188. Saddow SE, Schattner TE, Brown J, Grazulis L, Mahalingam K, Landis G, R. Bertke, Mitchel WC (2001). Effects of Substrate Surface Preparation on Chemical Vapor Deposition Growth of 4H-SiC Epitaxial Layers. Journal of Electronic Materials 30(3):228–234

Chapter 3
Silicon Carbide Electronics

One of the major benefits of silicon carbide for harsh environment microsystems is the ability to create high temperature electronics from a corrosion resistance base material. Because silicon carbide is a wide band semiconductor, it is more robust to high temperature excursions. But silicon carbide electronics requires the ability to create a substrate and thin-film layers that are high purity and can be doped in a controlled manner. The materials developments outlined in Chapter 2 lay the foundation for developing silicon carbide electronics. Besides being able to create doped, high-purity films, silicon carbide electronics requires a way to create localized doped regions in order to create specific transistor topologies as well as a metallization scheme for routing signals. This chapter will begin with a generalized process flow for creating silicon carbide electronics, followed by discussions on ion implantation doping and electrical contacts for silicon carbide. Then different electrical device topologies explored in silicon carbide will be described in the context of high power switching, high temperature amplifiers, and wireless communication.

3.1 General semiconductor process flow

A semiconductor process flow for SiC electronics fabrication is very similar to that of silicon and similar techniques are used. A run-down of the major process steps to create a SiC Junction gate Field-Effect Transistor (JFET) follows. Figure 3.1 schematically illustrates the process layers in a representative n-channel JFET fabrication process flow. But the processes are similar for other types of electronics devices or circuits. The n-channel JFET process begins with an n^+-doped substrate upon which epitaxial layers of p, n, and p^+ SiC are grown. If so desired, these epi-layers can be specified when ordering the SiC substrate. This defines the base doping levels used to create different features of the JFET. Next, a photolithographic process is used to first etch back the p^+ layer except to define the gate. This can be accomplished directly with photoresist as the etch mask or the pattern is first transferred from the photoresist pattern to a "hard mask" material such as nickel. The latter is typical for SiC. Using a "hard mask" allows finer feature resolution and minimizes shape changes due to mask erosion during the SiC etch. The mask material is cleaned from the wafer after the etch so as to not trap the mask material under subsequent depositions.

Another hard mask material, for instance silicon dioxide, is deposited and patterned to define the regions of doping using ion implantation. Ion implant will be discussed in more detail in the subsequent section. After the ion implant, the hard mask material is removed from the surface. This completes defining the doped regions of the device. Drawing parallels from silicon processing, it may be beneficial to follow this with a light nitrogen doping to reduce the series resistance at the gate-to- drain or gate-to-source, although this is omitted from the basic process flow in Figure 3.1. A thermal cycle to 1200 °C is used to activate the implanted dopant.

Next, another dielectric film is grown or deposited and again patterned to provide contact vias. Vias are needed to complete the routing of the interconnect layer to the source, gate, and drain of the JFET. Current device research may sometimes grow silicon dioxide on the SiC surface using a high temperature furnace; however, silicon dioxide films grown from SiC substrates tend to be of poor quality. This will be more fully explored in Section 3.2. As will be pointed out in more detail, alternate dielectrics are strong contenders for use with SiC. In particular sputtered α-AlN and LPCVD silicon nitride are both good insulators and have CTE relatively close to SiC, which is beneficial from a thermal stress perspective. In a research environment, the dielectric may simply be deposited, but no further processing to planarize the resulting surface is pursued (see Section 2.6, Planarization). This step is not technically necessary for research because lithographic resolution to increase device intensity is not critical when trying to make just a handful of functioning devices. As density and cost become important, planarization becomes crucial in order to optimize lithographic pattern size of subsequent steps. The dielectric is patterned to allow electrical access to the underlying transistor features.

The electrical contact metal is deposited and lithographically patterned. A range of metals are used, ranging from aluminum to nickel to platinum. After the photodefined mask material is removed, the most basic version of a SiC electrical device is defined and can be tested. If additional routing is not needed, another dielectric is deposited over the substrate and again patterned to provide electrical contact to the circuit. However, as devices begin to be connected, the process can be extended to provide a second metal routing layer following the same sequence of metal deposition, inter-metal dielectric deposition, and final dielectric passivation and patterning to provide bond wire access. As circuitry repeats, additional metal routing layers can be added. For example, commercial silicon semiconductor processes may have 3 to 7 metal routing layers. Making good electrical contact to SiC, especially for high temperature applications, has been a challenging area of research and will be discussed in Section 3.3.

Some special features may be added to this generic process flow depending on the particular device being created. For instance, to make an insulated gate transistor, a trench needs to be formed between the n^+ regions to form a physical way of separating source nodes. Also, most power devices operate using a substrate contact

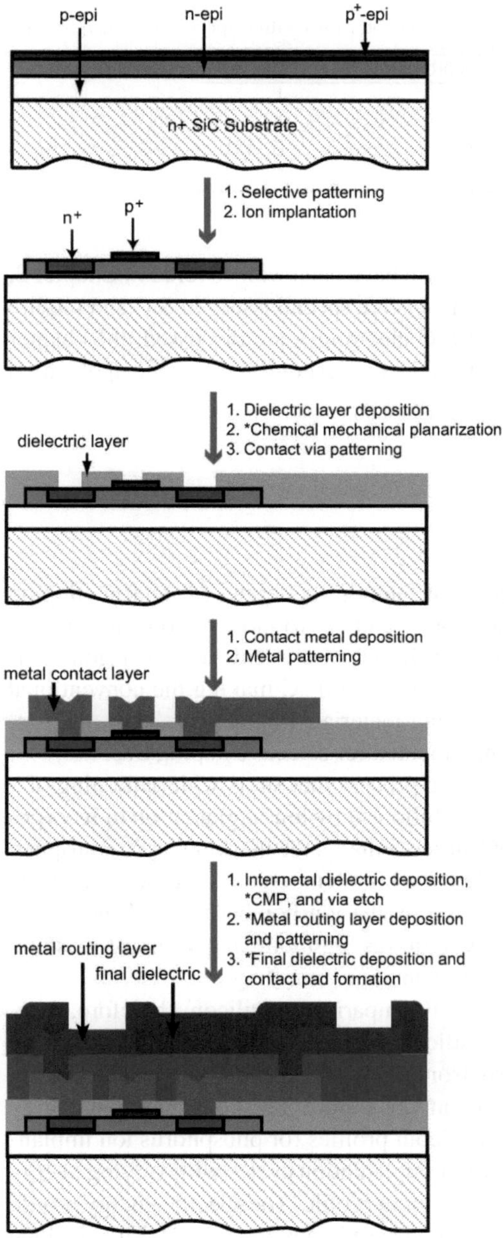

Fig. 3.1 Generic cross-section schematic process flow for a n-channel junction gate field-effect transistor with two metal layers.

as the cathode or collector node of the device, which requires an additional metal deposition and patterning on the backside of the wafer.

Table 3.1 Penetration depth of common donor and acceptor ions with respect to ion-energies [4]

Ion Type	Penetration Depth (μm)	Ion energy (MeV)
n^+ (donor)	2.3	4
p^+ (donor)	2.0	4
B+ (donor)	1.3	1
Al+ (donor)	0.9	1

Although etching and epitaxial film growth have been well covered in Chapter 2, ion implantation doping and electrical contacts are special issues in SiC and require additional detail. After these aspects of SiC electronics are discussed, the chapter will explore various transistor topologies in the context of three specific application areas.

3.2 Ion implantation doping

SiC device fabrication requires creation of laterally patterned p^+ and n^+ doping regions. In silicon technology, p^+ (acceptor) and n^+ (donor) regions are created using diffusion or ion implantation doping. In the case of SiC, the diffusion coefficients of most impurities are negligible below 1800 °C making the conventional diffusion techniques using standard masking materials impractical. Thus, the ion-implantation technique is generally implemented for selective doping SiC.

Selective ion implantation is a three step process that starts with patterning of a masking layer to define the doping region followed by ion implantation to incorporate donor or acceptor ions into the SiC lattice. Finally, the masking layer is stripped and the ion implanted sample is annealed at a high temperature in order to activate the dopant and heal lattice damage caused by the ion bombardment. When compared to ion-implantation of silicon, SiC requires high ion-energy, robust masking materials, and higher temperature post-implantation annealing.

The density of SiC is relatively high in comparison to silicon; therefore, it requires higher ion-energies compared to silicon in order to create an equivalent doping profile [1]. The energy range starts from a few tens of keV for shallow doping to as high as a few MeV for deep implant [2]. Figure 3.2 shows how ion energy affects the dopant distribution at various depth profiles for phosphorus ion implantation into 6H-SiC [3]. Typically, a few ion implantation cycles with varying energy are employed to obtain a uniform vertical doping profile. The specific energy profile used for fabricating a particular device can be based on many factors. They include device-architecture-imposed doping profile constraints as well as the type of the ion. With regards to the type of the ions, penetration depth is higher for the smaller ions than that of larger ions at the same ion energy. The predicted penetration depth of common donor and acceptor ions are presented in Table 3.1.

Generally, nitrogen and phosphorus ions are used in ion implantation of SiC to form the n-type region while aluminum and boron are used as the p-type dopant. The

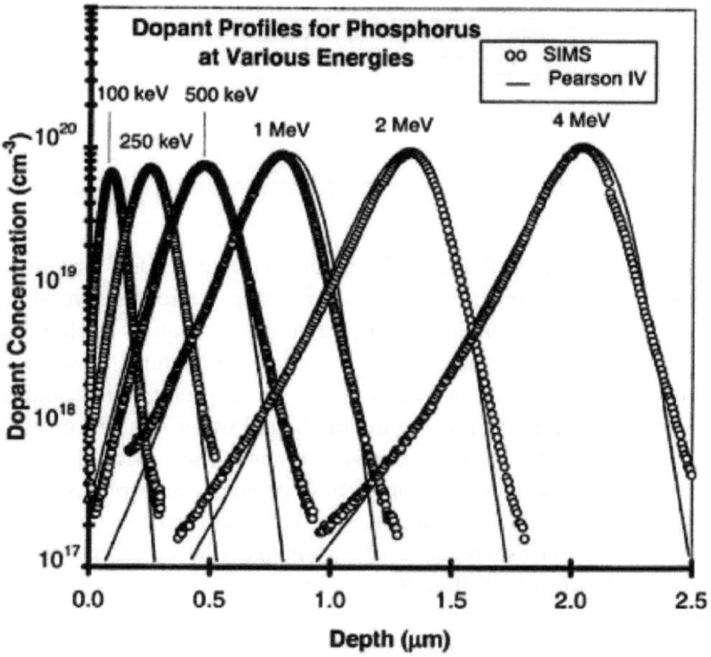

Fig. 3.2 [3] Experimental depth profile for phosphorus ion implantation into 6H-SiC at various energy (©Springer 1999), reprinted with permission.

shallowest ionization energy for these dopants for 4H-SiC and 6H-SiC are listed in Table 3.2 [5, 13]. The high ionization energy of these donors and acceptors in SiC limits the achievable carrier concentration in devices operating at room temperature. This also demands a higher concentration of dopant atoms to be implanted, causing greater lattice damage. Phosphorus has an advantage over nitrogen when high donor concentration is required because of its high solubility in SiC. Aluminum ion implantation is preferred for forming a low-resistance p-type region because the ionization energy of aluminum is shallower than that of boron; however, boron ion implantation is effective for ion implantation in deeper regions because of its higher penetration depth due to its low atomic mass. Efficient activation of p-type dopant in a wide bandgap semiconductor is relatively difficult. Nonetheless, p-type doping is routinely done because it is extremely important for many device applications.

As previously discussed, the higher ion energy and larger implantation dose required for SiC ion implantation causes increased lattice damage compared to silicon. To repair this ion-induced lattice damage and electrically activate the dopant by migrating implanted ions from interstitial to lattice sites, a post-implantation anneal is required. The annealing temperature varies with ion dose and ion energy because these factors are directly related to the crystal damage. The annealing tem-

Table 3.2 The shallowest ionization energy for dopants in 4H-SiC and 6H-SiC [5, 13].

Dopant	Ionization energy (meV) for 4H-SiC	Ion energy (MeV) for 6H-SiC
Nitrogen	45	85
Phosporus	80	80
Boron	285	300
Aluminum	190	240

perature typically ranges from 1200-1800 °C, higher than that needed for silicon. Even though post-implantation annealing is a required step, it creates several unwanted problems. First, this very high temperature annealing causes preferential silicon evaporation from the SiC surface leading to surface roughening. However, it can be mitigated by maintaining silicon overpressure (silicon-rich ambient) during the anneal. This is typically done by placing the implanted wafer in a SiC crucible with SiC powder inside or flowing silane during the annealing process [7]. Second, if ion implantation causes the formation of fully or nearly amorphous SiC, it is difficult to fully restore the crystal structure. The very high temperature conditions required for anneal lead to segregation of carbon and silicon leading to non-stoichiometric regions and formation of multiple polytypes. These factors adversely affect the electrical characteristics of the doped regions making them unsuitable for device fabrication. Therefore, it is desired to keep lattice damage to a minimal level during the ion implantation process, which in turn allows lowering the post-implant anneal temperature. For this reason, ion implantation of SiC is generally carried out at elevated temperatures (500-1000 °C) [1]. The elevated-temperature assists some crystal restructuring during the implantation process and reduces the lattice damage and segregation of displaced silicon and carbon atoms [8, 9].

The high energy and the high temperature conditions of SiC ion implantation impose extra requirements on the masking materials. First, the high ion-energy requirement demands that masking materials be thicker or denser than masking materials used for obtaining a similar doping profile in silicon. The elevated temperature during implantation requires a mask layer that withstands these conditions. Currently, either a thick silicon dioxide layer or refractory metal such as nickel is used as the robust masking material [1].

3.3 Contacts to SiC

SiC microsystems require metal to SiC electrical contacts for both MEMS and electronics. Both ohmic and Schottky contacts are necessary to transfer signals in and out from sense electronics, MEMS, and communication circuitry. Schottky contacts are particularly required for SiC electronics that provide switching and rectification. To fully exploit SiC for harsh environment microsystems, electrical contacts between SiC and metal must be reliable for operation in extreme conditions such as high temperature and high power. Even though several promising advances have been made in forming both ohmic and Schottky contacts, reliability and long-term stability of metal-semiconductor contacts are key limiting factors that hinder operating SiC-based electronics in extreme conditions.

3.3.1 Ohmic contacts

Electrical connections between a metal and a semiconductor that have a linear current-voltage (I-V) response curve with very low specific contact resistance are called ohmic contacts. In order to create these conditions, negligible Schottky barrier height (SBH) is needed. Wide-bandgap semiconductors have relatively large SBH when compared to narrow-bandgap semiconductors due to the larger work function difference between the metal and semiconductor. Each SiC polytype has a slightly difference band gap height. The work function of different metals also vary. Therefore, both SiC polytype and contact metal influences the SBH. There are many other parameters that affect the SBH as well. Dopant concentration, type of dopant (n or p), pre-metal-deposition surface preparation, metal deposition process, and post-deposition annealing conditions, all have an impact on the electrical characteristics of the SiC-metal interface [10, 11, 12].

In general, almost every metal-SiC interface exhibits Schottky (rectifying) behavior when the metal is in the as-deposited state [13]. However, certain process steps such as post-deposition annealing can be implemented to convert rectifying contacts to ohmic contacts. Annealing typically creates a reaction layer between the metal and SiC, which helps to reduce the SBH. The reaction layer consists of metal-silicide, -carbide, or both. Post-deposition annealing is typically performed between 950-1100 °C. Optimizing annealing conditions is highly critical for each different combination of metal, SiC polytype, and dopant in order to obtain the lowest contact resistance possible. Figure 3.3 shows the I-V Characteristics of a TiW/3C-SiC contact as a function of annealing temperature and heating method: vacuum oven or rapid thermal annealing (RTA) [15].

Dopant concentration and type both heavily influence the resulting contact resistance. For instance, specific contact resistance values for nickel on n-type SiC range between 10^{-2} and 10^{-6} Ωcm^2 for the same annealing and deposition conditions. Lower contact resistance is achieved if the dopant concentration exceeds 10^{19} cm^{-3} [12]. For most metals, p-type SiC exhibits higher SBH at the metal-semiconductor interface and requires higher annealing temperatures in order to form ohmic contacts. Furthermore, lowering contact resistance when forming a contact on p-type SiC requires a higher dopant concentration than that the equivalent n-type SiC [12]. Typically, ohmic contacts to p-type SiC contain aluminum because aluminum enhance the p-type concentration at the SiC surface, which in turn reduces the SBH.

Many different metals have been investigated. Nickel is one of the widely used metals for creating ohmic contacts to n-type SiC. Nickel contacts to n-type SiC achieve low specific contact resistance and exhibit excellent electrical and physical

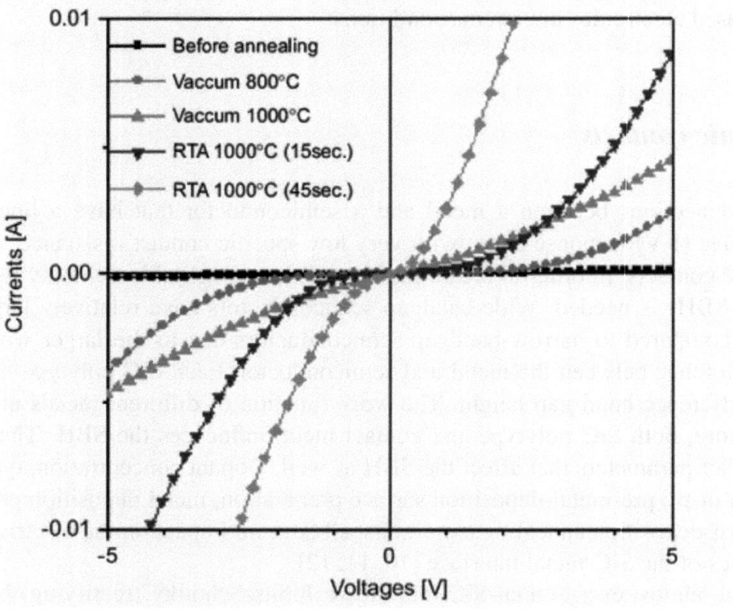

Fig. 3.3 [15] I-V Characteristics of TiW/3C-SiC contact depending on annealing temperature and conditions: vacuum oven or rapid thermal annealing (©Springer 2008). Reprinted with permission.

stability, making them a good choice for long-term operation of high power devices. Additionally, many refractory metals such as titanium, molybdenum, tantalum, and chromium as well as multi-layered structures have been successfully implemented. Multi-layered stacks include titanium/tungsten, titanium/tungsten/nickel, and titanium/titanium silicide/platinum (Ti/$TaSi_2$/Pt) [12, 16]. For ohmic contact to p-type SiC, generally, aluminum and aluminum-based alloys such as AlTi, AlTiW are used [11, 12].

Most ohmic contacts developed to date provide sufficiently low ($< 10^{-6}$) contact resistance needed for device operation. These contacts have been experimentally validated for long-term operation up to 300 °C [13]. However, some harsh environment SiC microsystem applications require contacts that can reliably withstand 500-600 °C ambient. To date, the reliability of contacts is the limiting factor for high temperature operation of SiC devices. Many contact metals tend to further react with SiC during high temperature operation, resulting in an increased reaction layer, which eventually impacts contact performance. Furthermore, oxidation and electromigration also degrade the electrical quality of the metal-SiC interface.

Both single- and multi-layer metallization schemes have been investigated for creating stabile ohmic contacts for high temperature operation. Single-layer metal contacts exhibit increased specific contact resistance when operating temperatures exceed 300 °C. Multi-layer metallization schemes show greater promise. Recent work also investigates creating a nanocrystalline carbon interfacial layer on the SiC surface before depositing a Pt contact metal [14]. This contact is tested up to 540 °C with negligible increase in contact resistance. One of the notable multi-layer schemes, Ti/$TaSi_2$//Pt, developed at NASA Glenn Research Center, has shown to operate over 1000 hours in air at 600 °C without significant changes to contact resistance[16]. Later, this Ti/$TaSi_2$/Pt metallization was successfully implemented in the fabrication of SiC JFET devices that have demonstrated stable operation for over 10,000 hours at 500 °C [35]. Despite these successes, many aspects of forming metal contacts to SiC have to be investigated. These include identifying thermodynamic, reaction kinetic and dominant failure mechanisms of the metal-SiC interface at elevated temperatures in oxidative environments.

3.3.2 Schottky contacts

Schottky contacts are vital for many SiC devices including the widely used high-frequency metal-semiconductor field-effect transistor (MESFET) and fast-switching rectifiers. The larger SBH and thermal stability of the contact are key parameters for high temperature operation and low power loss. The SBH depends on many parameters, including the work function of the metal. According to Schottky-Mott theory, high work function metals are required for larger SBH. Pt, Ni, Au, and Pd are preferred Schottky contacts to SiC because of their relatively high work function [18, 19]. The pre-metal deposition surface preparation also can cause dramatic changes in the resulting SBH. For instance, Teraji *et al.* has shown drastic SBH changes for the same contact metal, titanium, when two surface cleaning procedures are used [20]. Furthermore, SBH strongly depends on chemical reactions at the metal-SiC interface. Therefore, interfacial physics and chemistry aspects must both be considered in order to fabricate reliable Schottky contacts.

Although almost all metal-SiC contacts exhibit Schottky (rectifying) behavior in the as-deposited state SiC, they tend to form either a silicide or carbide interlayer, or possibly both, at high temperatures, reducing the SBH. This leads to an increase in reverse-bias thermionic leakage of carriers over the junction barrier. State-of-the art SiC Schottky diodes typically are limited to operation below 400 °C because of the reverse-bias (off-state) leakage current [13]. Similar to ohmic contacts, there are few examples of thermally stable Schottky contacts at high operating temperatures. For instance, some limited success in creating thermally stable contacts has been reported using tungsten carbide deposited at high temperature. Tungsten carbide deposited on p-type 6H-SiC exhibits a high rectification ratio with a low reverse current density (6.1×10^{-5} A cm^{-2}, -10 V) up to 500 °C [21]. Recent reports shows refractory metal borides are also very promising for forming stable high temperature

Schottky contacts to SiC [22]. However, more research focusing on this issue is needed.

3.4 Electronics for high power applications

A special application space for electronics is regulating high power devices. Vacuum tube components were replaced with silicon transistors in the 1950s. Specialized topologies requiring non-standard microfabrication compared to microprocessor and memory are often utilized (substrate trenching, multi-level doping profiles, and backside contacts). The added cost from not using the most standardized processes is well compensated by the important applications enabled by these high power switches and rectifiers. Power applications are plotted in Figures 3.4 and 3.5 [23]. Figure 3.4 plots applications as a function of frequency and power rating. As can be seen by the graph, high frequency applications tend to have lower power handling demands compared to low frequency applications. Figure 3.5 makes a similar distinction, but this time between voltage and current handling capabilities. The device topology differs depending on operating frequency, required voltage and current capability, or blocking capability. For instance, applications below 100 V are well served by silicon Schottky diodes. However, above this voltage, bipolar topologies are used [23].

Silicon carbide has been investigated for high power switching and rectifying applications because it is a wide-band semiconductor. This enables extending the higher performance of unipolar devices to operating voltages as high as 5000 V instead of needing to switch to bipolar device topologies [23]. As will be discussed in more detail, the higher blocking characteristic of the silicon bipolar device comes at the cost of switching transients. These transients are inefficient periods compared to ideal behavior, resulting in power loss and an effective degradation in switch frequency. Furthermore, high power switching generates significant heat that needs to be dissipated. Silicon carbide has nearly three times the thermal conductivity as silicon, which aids in heat dissipation. But being a wide-band semiconductor also allows silicon carbide devices to operate to higher temperatures without thermally-induced leakage due to intrinsic carrier concentration variation with temperature. The major device topologies and design tradeoffs in the context of high power handling will be discussed in subsequent sections.

3.4.1 Schottky Rectifier

The Schottky rectifier is a popular unipolar device because its basic fabrication requirements are simple, it offers fast switching speed, and it has a relatively low on-state resistance. The one-dimensional representation of the structure is shown in Figure 3.6. The basic Schottky rectifier is created by depositing a metal onto the

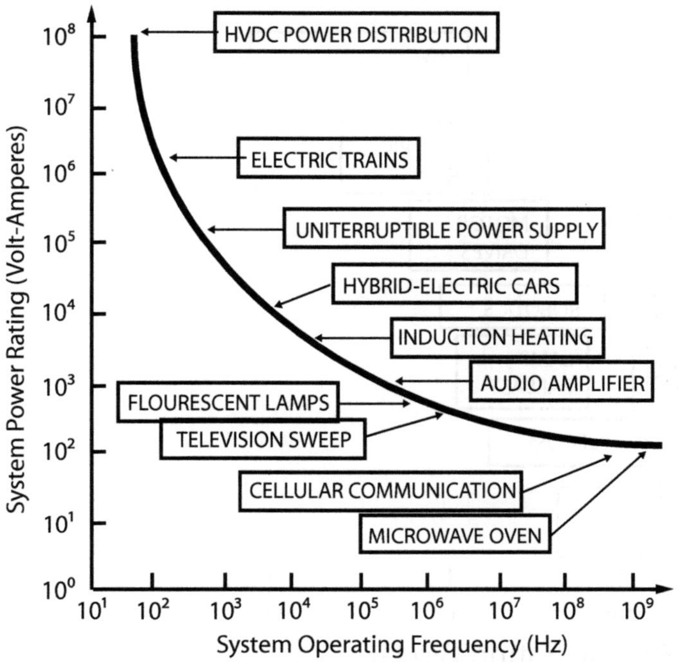

Fig. 3.4 [23] High power applications versus operating frequency and power handling requirements (©Springer 2009). Reprinted with permission.

substrate surface. The metal becomes the anode region while the backside of the wafer is the cathode. The applied voltage between the anode and cathode creates a drift region within the substrate, next to the anode. By applying a negative bias to the cathode, current flow occurs by electrons migrating from the metal-semiconductor interface through the substrate. While in the on-state, current is sustained using majority carriers. This means few minority carriers build up within the drift region. This enables rapid switching to the off-state by establishing a depletion region inside the drift region. Silicon devices of this type can support blocking voltages of 100 V.

The Schottky rectifier can obtain a larger barrier height in SiC because of the increased band-gap compared to silicon. This allows supporting even higher blocking voltages. The majority of applications shown in Figure 3.4 require fast switching, low on-resistance, and blocking voltages of 500 V or less. SiC Schottky rectifiers can be used to span a large region of the application space since it is possible to extend the breakdown voltage to 3000 V if using SiC [23].

The drawback of the Schottky rectifier is large off-state (reverse bias) leakage current. Because of the present state of achievable n^+ substrate doping for SiC, the substrate resistance becomes a concern for power dissipation: Si can be as low as 1 mΩ-cm, compared to SiC, which is typically 20 mΩ-cm. The barrier height and saturation current is also a strong function of operating temperature in silicon,

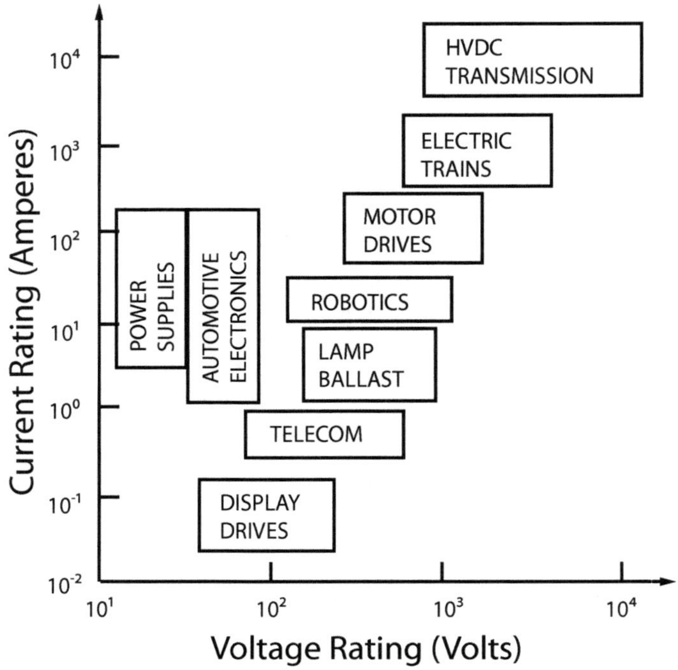

Fig. 3.5 [23] High power applications versus voltage and current requirements (©Springer 2009). Reprinted with permission.

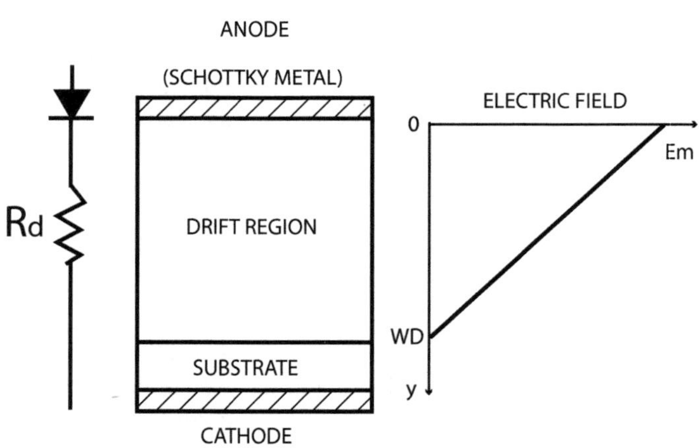

Fig. 3.6 [23] Electrical field distribution in a Schottky rectifier (©Springer 2009). Reprinted with permission.

which increases both the on-state resistance and off-state leakage current. However, the net impact of increased temperature on a SiC Schottky rectifier is smaller. SiC Schottky rectifiers can also leverage the larger barrier height to create a rectifier that is designed to maintain an acceptable power dissipation in the reverse blocking

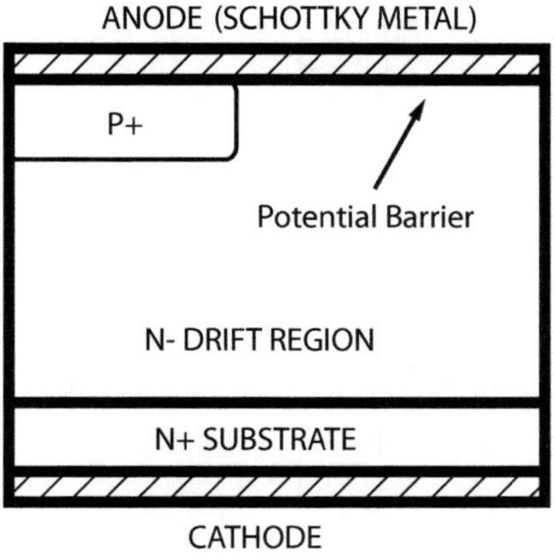

Fig. 3.7 [23] Schematic silicon carbide Junction Barrier controlled Schottky (JBS) rectifier structure (©Springer 2009). Reprinted with permission.

mode even though the leakage current increases more significantly with increased reverse voltage compared to silicon. For example, a particular 3kV 4H-SiC Schottky rectifier at room temperature exhibits less than 1 W/cm^2 in the off-state compared to 100 W/cm^2 of dissipation in the on-state [23].

To combat the leakage current problem with the Schottky rectifier configuration, various approaches to shield the metal-semiconductor interface from directly sustaining a large electric field, such as utilizing a p-n junction or introducing a second metal with an even larger barrier height, have been introduced. These two methods are known as the Junction Barrier controlled Schottky (JBS) and Trench Schottky Barrier controlled Schottky (TSBS) rectifier schemes, respectively. These methods will be discussed in more detail next.

3.4.2 JBS Rectifier

Schottky rectifiers balance on-state power loss with reverse blocking power loss through reduction in Schottky barrier height until the reverse blocking power loss dominates. This in turn reduces the maximum operating temperature. However, as was discussed in the previous section, this reverse blocking power loss occurs because of the high electric field across the metal-semiconductor interface. The Junction-Barrier controlled Schottky (JBS) rectifier attempts to prevent creating a high electric field when in the off-state by injecting closely spaced p$^+$ regions, creating local p-n diodes, directly under the anode (Figure 3.7). These p$^+$ regions create a potential barrier to shield the contact. A smaller separation and larger junction depth improves the electric field reduction. However, the p$^+$ regions impede conduction during the on-state. Hence, the spacing must be optimized to meet both on-state and off-state performance specifications.

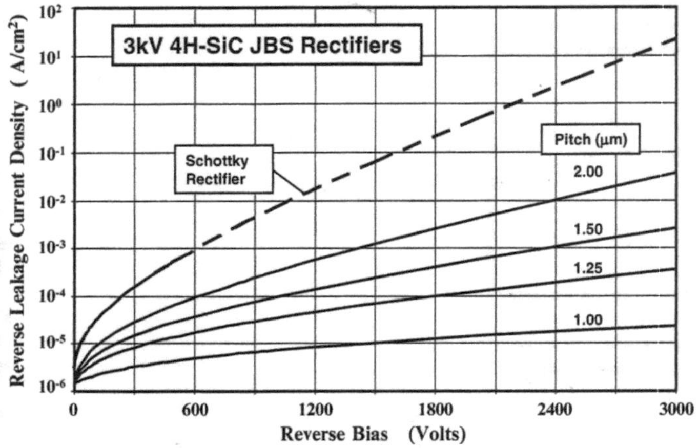

Fig. 3.8 [23] Leakage current suppression in a JBS rectifier as a function of p-region pitch ((©Springer 2009). Reprinted with permission.

Since the reverse blocking leakage current is more significantly influenced by a high voltage because of the more efficient generation of current through thermionic field emission (tunneling), the JBS rectifier structure can improve the performance of silicon carbide devices significantly. In SiC devices, the p-type dopants are introduced using ion-implantation, with the mask spacing dictated by the space needed to make the Schottky contact. The same metal layer used to make the Schottky contact to the n- region is typically used to make contact to the p^+ region for processing convenience [23].

The influence of the pitch of the p^+ region on reverse blocking leakage current is shown in Figure 3.8. Reducing the pitch down to 1.5 to 1.25 μm reduces the electric field at the metal-semiconductor interface by about half. However, this results in a 10^5 reduction in leakage current. This pitch range also demonstrates excellent on-state conduction when the p^+ region depth is 0.5 μm and is paired with a 20 μm drift region in order to support blocking 3kV in the off-state. Further decreasing the pitch increases the on-state power loss compared to a standard SiC Schottky diode of the same dimensions [23]. It should be noted that this pitch is fortunate from a practical perspective. Because of the current SiC wafer sizes (100mm diameter or less), equipment for handling substrates in this size range can have difficulty producing highly-repeatable, sub-micron dimensions.

Although this approach is popular for SiC rectifiers because it creates a dramatic reduction in off-state leakage current and small increase in resistive loss in the on-state without the need to resort to a bipolar configuration. A drawback though is the additional ion-implantation requires an activation anneal and creates additional lattice damage. This results in the need for very high temperature annealing (approximately 1600 °C). Thus, another method to create a similar shielding of the Schottky contact will be explored next that does not rely on an additional ion-implantation.

3.4.3 TSBS Rectifier

The drawback of the Schottky rectifier for high voltage applications is the high leakage current in the off-state due to thermionic emission from the metal-semiconductor interface when it is supporting a large electric field. Another method of shielding the Schottky contact from having to support a large electric field is to place a second high barrier Schottky metal within a vertically-walled trench. This device topology is known as the Trench-Schottky-Barrier controlled Schottky barrier (TSBS) rectifier.

The device topology essentially replaces the p^+ doped region buried under the anode with a physical region refilled with a metal that has a higher barrier potential than the anode metal. For example, a 4H-SiC device with blocking voltages of 300 V that uses titanium as the main Schottky contact and nickel in the trenches has been successfully demonstrated [24]. This additional processing complexity compared to ion-implantation removes the need for a post-implantation anneal at 1600 °C, which tends to roughen the SiC surface, degrading the anode-semiconductor contact. The net effect of the TSBS rectifier topology is similar to the JBS rectifier and similar design implications apply. Deeper, narrower trenches with close pitches on the order of 1 to 1.5 μm provide better shielding of the anode metal to semiconductor interface, reducing the resulting leakage current in the off-state while maintaining nearly equivalent on-state performance. This comes at a compromise in breakdown voltage characteristics, reducing it to approximately 80 percent of the ideal parallel plate due to edge termination [23]. However, this is equivalent to the JBS rectifier as well.

In the Si rectifier, the on-state voltage drop versus the reverse leakage current are very similar for a JBS configuration versus a TSBS configuration designed for 50 V. However, performing the same analysis for a 3 kV SiC rectifier, the TSBS configuration exhibits approximately 0.2 V lower on-state voltage drop compared to the JBS configuration (Figure 3.9). This is attributed to a larger depletion width under on-state operation in the JSB rectifier structure in 4H-SiC because the built in potential for the p-n junction is much larger than the contact potential for the metal in the trenches of the TSBS rectifier [23].

3.4.4 P-i-N Rectifier

The TSBS and JBS topologies work well for certain voltage ranges. However, for motor control applications, much higher voltage blocking and current carrying capability are required. In these applications, the on-state voltage drop of these topologies becomes significant. For these applications, silicon p-i-n rectifiers were developed. The i-region in this case is a low doped n- region between a highly doped p^+ and n^+ region, essentially silicon at its intrinsic carrier doping level. When very large voltages are applied to this structure, conduction through the intrinsically doped region is dominated by injection of minority carriers.

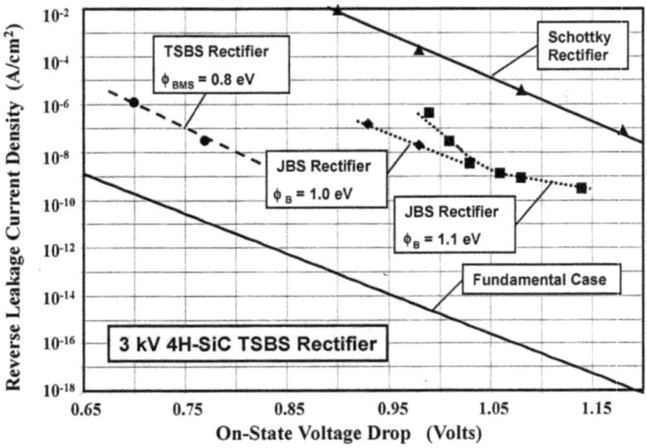

Fig. 3.9 [23] Trade-off curve for a 3 kV 4H-SiC TSBS rectifier compared to a JBS and Schottky rectifier (©Springer 2009). Reprinted with permission.

Because the device physics utilized in this structure injects a large number of free carriers into the drift region during on-state current flow, switching the p-i-n rectifier to the off-state to the point where the rectifier can again support a high voltage requires removing these carriers. Figure 3.10 presents simulation results of free carrier density versus distance into the drift region. Each contour represents time after the rectifier is switched to the off-state. Removing free carriers results in effectively slowing down switching from the on-state to the off-state compared to the previously discussed rectifiers. This is known as reverse recovery. This switching time equates to a power loss during switching. Additionally, this loads the other components of the circuit, making it necessary to increase their breakdown voltages as well.

The SiC p-i-n rectifier has a much narrower drift region for the same breakdown voltage capability compared to the silicon equivalent. This enhances the switching speed characteristics of a SiC p-i-n rectifier. However, the same large bandgap that enables the high breakdown voltage increases the on-state voltage drop by a factor of four compared to silicon. The design trade-off between speed and on-state resistance favors SiC p-i-n rectifiers over Si p-i-n rectifiers or SiC Schottky diodes when the voltage ratings exceed 10 kV.

3.4.5 MPS Rectifier

The p-i-n structure has been shown to be effective for very high power applications; however, it suffers from slow switching speed because it relies on minority carriers. These minority carriers must be dissipated before full reverse bias block-

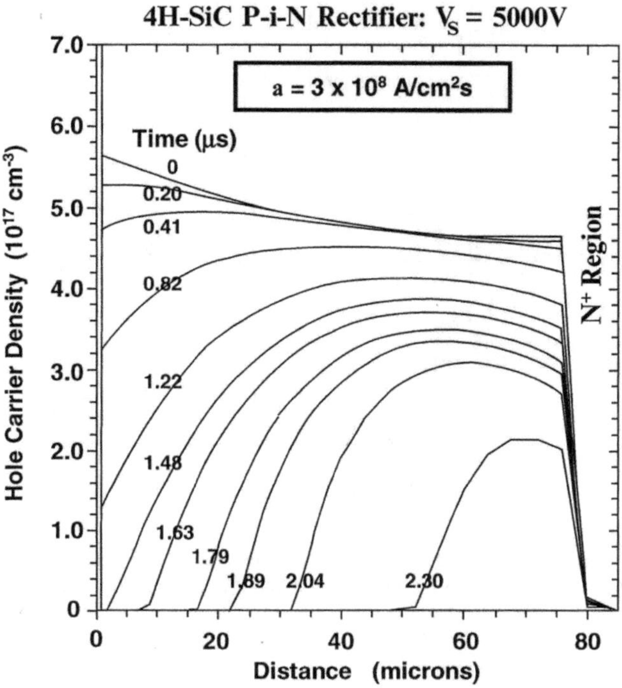

Fig. 3.10 [23] Carrier distribution in a 10 kV 4H-SiC p-i-n rectifier during the reverse recovery transient (©Springer 2009). Reprinted with permission.

ing is achieved. This delay limits the maximum switching speed and also creates a switching power loss during the on-to-off transient. In order to reduce the transient, a structure that reduces the conductivity during the on-to-off transition period is needed. Although the topology can be schematically represented identically to the JBS rectifier (Figure 3.7), the Merged Physics Structure (MPS) rectifier substitutes the n^- drift region of the JBS rectifier with a drift region that is operating as an intrinsic semiconductor, identical to the p-i-n drift region. The MPS rectifier can also be seen as putting a p-i-n rectifier in close proximity to a Schottky rectifier. However, assuming the structures do not interact would lead to the conclusion that the MPS structure would yield the worst performance of each topology. But in fact MPS works because of an interaction between these structures. Like the JBS and TSBS rectifier topologies, the pitch of interleaving the p-i-n rectifier and Schottky rectifier influences the overall device performance. With careful design, the MPS structure will outperform the p-i-n structure for a given current handling and blocking voltage design point.

The Schottky rectifier portion of the MPS structure has a lower potential barrier than the p-n junction in the p-i-n region, which aids in reducing on-state resistance (forward voltage drop) compared to a standard p-i-n (Figure 3.11). This reduced bar-

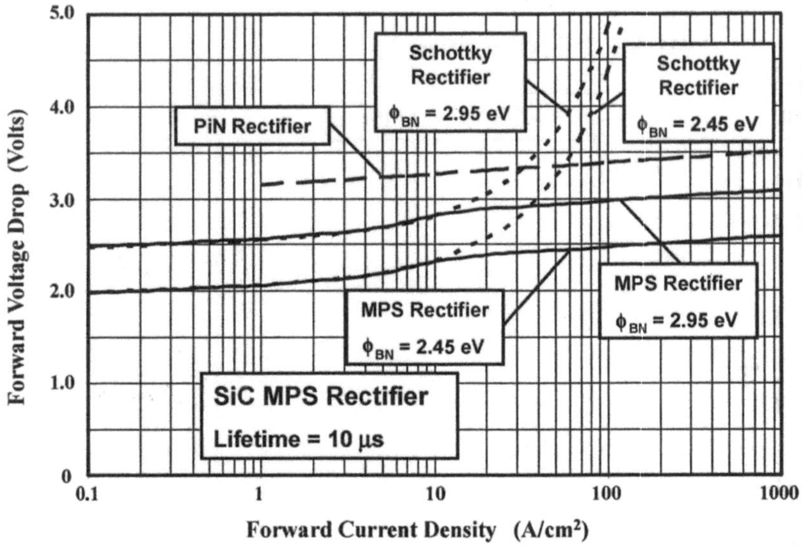

Fig. 3.11 [23] On-state characteristics of 10 kV SiC MPS rectifiers of different barrier heights as well as comparative curves for equivalent p-i-n and Schottky rectifier designs (©Springer 2009). Reprinted with permission.

rier path results in up to a 10 fold decrease in stored energy in the rectifier in the on-state. In the reverse blocking mode, the Schottky rectifier region causes larger leakage current under high applied voltages. This is combated by reducing the Schottky contact width significantly compared to the p-i-n region. However, switching speed is improved because the Schottky contact portion of the MPS enables supporting some voltage immediately when the rectifier is commanded to the off-state. During the transient while the injected minority carriers associated with the p-i-n structure are dissipated, the Schottky rectifier portion reduces unwanted power dissipation. This results in a 10 kV SiC MPS rectifier having approximately half the peak reverse recovery current and half the overall reverse recovery time compared to an equivalent p-i-n rectifier according to numerical simulations [23].

3.4.6 BJT

As was discussed in Section 3.4.1, unipolar rectifiers are preferred because they exhibit acceptable power dissipation and faster switching speed for lower voltage applications — this limit is 300 V for Si and 3000 V for SiC. Hence, SiC is seen as providing an overall performance improvement compared to silicon-based technologies between 300 V and 3 kV. However, exceeding 3 kV, bipolar SiC topologies need to be considered.

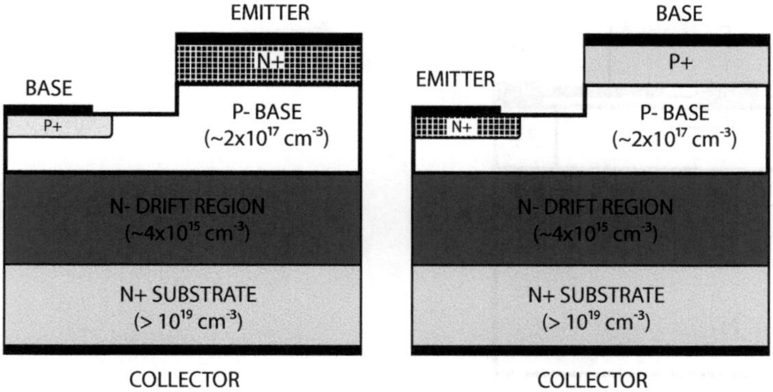

Fig. 3.12 [26] Schematic cross-section of (left) an epitaxial and (right) implanted emitter SiC bipolar junction transistor (©Springer 2004). Reprinted with permission.

The Bipolar Junction Transistor (BJT) is one such topology (Figure 3.12). Bipolar devices typically exhibit lower breakdown voltages compared to their p-n junction equivalents due to current gain; however, the *npn* configuration has been shown to be more rugged, so is more common [26]. As was discussed for the SiC JBS configuration, an ion-implanted p^+ region requires a very high temperature anneal that is best to avoid. Hence, the majority of SiC *npn* BJT topologies rely on a non-planar structure that utilizes an epitaxially grown layer to create the p^+ base region and create the n^+ emitter region through ion-implantation.

A BJT designed for high power applications will have a lightly-doped collector region and greater substrate thickness than low-voltage BJTs in order to prevent punch-through. This lightly-doped collector region reduces output current and slows down switching speed for the same reasons discussed for the p-i-n rectifier. Along with current crowding effects, the realized current gain can be significantly degraded. Hence, the power BJT can be fabricated in a two-stage configuration, referred to as a Darlington configuration, which results in an effective current gain that is the square of the individual gain stages [26]. This lightly-doped region also results in a quasi-saturation current response at high voltages. Furthermore, using an ion-implanted emitter region results in degraded current gain compared to devices that use an epitaxially grown n^+ emitter and perform a p^+ ion-implantation for the base region. This is presumably due to degradation of the SiC surface at the contact region. As with other topologies that rely on minority carrier injection during the on-state to reduce the voltage drop, the BJT does not switch to the off-state quickly. Hence, BJT rectifiers are not good choices when power cycling above 10-20 kHz is required for a given application. The BJT is also a current controlled device, which is less favorable in terms of integration with other circuitry than a voltage controlled topology.

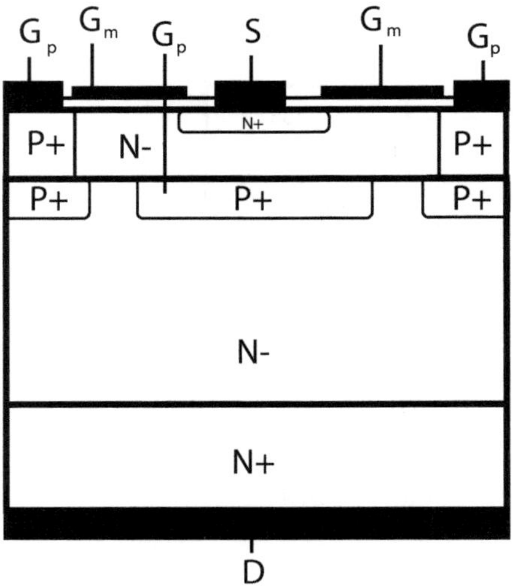

Fig. 3.13 [26] Schematic cross-section of a 4H-SiC SIAFET structure (©Springer 2004). Reprinted with permission.

3.4.7 SIAFET

Figure 3.13 is a schematic cross-section of a Static Induction Injected Accumulated Field Effect Transistor (SIAFET), also known as a bipolar-mode JFET. This structure integrates a JFET with a Metal-Oxide-Semiconductor Field-Effect Transistor (MOSFET) structure such that during the on-state the junction gate is over-driven beyond the turn-on voltage. By forward biasing the gate junction, minority carrier injection from the gate improves the conductivity of the channel and drift regions.

The SIAFET configuration has been experimentally demonstrated up to 6 kV using 4H-SiC. This device consists of a lateral accumulation MOS channel and a vertical JFET channel. By forward biasing the junction gate, the devices operate in a bipolar mode. Bipolar operation of a 2 kV SIAFET rectifier reduces the on-resistance to 172 mΩ-cm^2 compared to 1200 mΩ-cm^2 for unipolar operation [27]. Even higher blocking voltage designs (6.1 kV) still exhibit acceptable on-resistance, 732 mΩ-cm^2, when operated in bipolar mode.

3.4.8 Power MOSFET

The Power MOSFET is of interest because of it has significantly higher switching speed than other topologies and is voltage controlled, which is easier to implement. Furthermore, the on-state current exhibits a negative temperature coefficient, which enables parallel device configurations for additional current carrying capability. The Power MOSFET was first successfully developed using SiC in what is referred to as

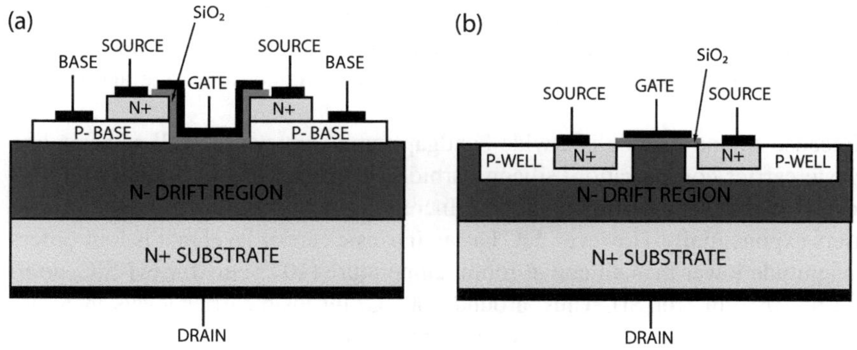

Fig. 3.14 [31] Schematic cross-sections of (a) UMOS and (b) DMOS power transistors in SiC (©Springer 2004). Reprinted with permission.

the U-shaped groove MOSFET (UMOSFET) [28]. This configuration can be made entirely from epitaxially grown layers to define the different doping regions. This in turn requires a trench be cut through the n^+ and p base region. The gate is formed by first oxidizing the trench and then coating it with the gate metal (Figure 3.14a). The reactive ion etch surface of the trench is rough and the oxide grown on SiC is of poor quality compared to thermally-grown oxide on silicon. Additionally, the growth rate is different on a Si-face surface versus a C-face surface of SiC, resulting in a thinner oxide at the base of the trench. The trench geometry also causes field crowding at the corners, causing repeated high fields being sustained by the trench oxide, which degrades device lifetime.

A more elegant solution, known as the Double-diffusion MOSFET (DMOSFET), was developed in 1998 [29]. This more planar structure eliminates the concern of poor oxide growth uniformity (Figure 3.14b). However, it added the need for ion-implantation to create the p base region and n^+ wells. This requires high-temperature annealing, which can be as high as 1600 °C and causes surface state degradation due to preferential evaporation of Si at these temperatures.

The SiC DMOSFET has been demonstrated as a 2.4 kV blocking device that operates in a normally-off condition. Up to 10 A current-carrying capability and 42 $m\Omega\text{-cm}^2$ on-state resistance has been experimentally demonstrated using a 3.3 by 3.3 mm sized device [30]. Devices able to block up to 1 kV have also been shown to operate up to 100 kHz. The interested reader is directed to [31] for details on design of a DMOSFET device.

3.5 Sensor interface electronics for high temperature applications

SiC electronics promise to open up additional application areas by potentially allowing reliable operation between 450–600 °C. Additionally, being able to operate at high temperatures would allow more conventional applications in the realm of 300 °C, such as for automotive engine monitoring, without the need for an active cooling system for the electronics. This significantly reduces the size and cost of such systems.

Because silicon carbide is a wide bandgap semiconductor as well as has a low intrinsic carrier concentration, silicon carbide electronics are well suited for significantly higher temperature operation. Increasing temperature increases intrinsic carriers exponentially. However, SiC has an intrinsic carrier level that is four orders of magnitude lower than silicon at room temperature (10^{-6} cm^3 for 6H-SiC, compared to 10^{10} cm^3 for Si). Thus, around 300 °C, intrinsic carrier levels in Si are approaching dopant levels for low doped regions, resulting in uncontrolled device behavior. Silicon carbide also has a very low rate of diffusion compared to silicon. Hence, dopant diffusion in silicon carbide electronics is not a concern for operating temperatures below 600 °C,whereas silicon circuitry, even when built on a SOI substrate, is limited to operating below 300 °C to avoid diffusion degradation of the electronics. This section focuses on the design of sensor interface electronics in the context of high temperature operation.

3.5.1 MOSFET and MESFET considerations

SiC MOSFET and Metal-Semiconductor Field-Effect Transistor (MESFET) device topologies are prevalent in high-frequency switching, moderate voltage blocking power electronics (Section 3.4.8) and RF communication components (Section 3.6). These topologies are desired for applications requiring fast switching speeds. However, they are not well suited for high temperature operation with the current state of dielectric processing for transistor applications. MOSFETs require a much higher quality silicon dioxide layer than other topologies. SiC does not grow a high quality thermal oxide, and this oxide further degrades with increased temperature [13]. This for instance leads to -14 mV/°C degradation in V_T above 200 °C [36]. Alternate dielectrics are an area of research that may open up the use of MOSFET structures for high-temperature operation. Deposited silicon dioxide for instance may be a workable alternative [37], but as with all deposited films, thickness uniformity comparable to grown oxide is a challenge. Switching instead to a deposited dielectric opens up a wide range of material options. Other materials that also have better CTE matching or better electrical stability with temperature may be possible. Additional research is needed in this area. Although there have been early reports of 500 °C operation of MESFET devices in the past [38], MESFET technology suffers from gate-to-channel leakage currents from Schottky diodes, limiting usable operation to around 400 °C [13]. Hence, these device topologies require further development if they are to be used for high temperature sense interface electronics.

3.5.2 JFET

The SiC JFET device is a promising device for MEMS sensor interface electronics. As was described in Section 3.1, the basic device can be fabricated in a non-planar configuration primarily from epitaxially grown layers with the exception of the n^+ source and drain regions, removing the need for very high temperature annealing. Furthermore, this structure does not rely on a high quality dielectric layer for proper operation. This topology is also a high-impedance input structure. This is well suited as a MEMS sensor interface, which primarily relies on detecting changes in capacitance through driving a sinusoidal signal onto the sense capacitor plates. This requires an amplifier with a high-impedance input.

SiC electronics are still primarily based on 4H- and 6H-SiC polytypes in part due to availability of quality substrates. 4H-SiC has a wider bandgap and higher electron mobility at room temperature compared to 6H-SiC. However, 4H-SiC does not perform as well at higher temperatures. There are more transistor designs that operate at very high temperatures using 6H-SiC [32, 33]. This is because the mobility drops off more rapidly with increasing temperature in both cases, but donor levels are lower in 4H-SiC.

A single-stage, single-ended transimpedance amplifier utilizing a JFET configuration has recently been demonstrated in 6H-SiC [34]. Devices were fabricated on 2 inch diameter 6H-SiC substrates. Since hole mobility is always less than electron mobility at any temperature, this device uses a n-channel. Thus, an aluminum doped p-type substrate is used. The device was formed primarily by etching back epitaxially grown layers to form a non-planar device structure (Figure 3.15). This limits the need for only one n^+ ion-implant step. However, in reality multiple doses at different energies were used to create a box profile. The post-implantation anneal was performed at 1200 °C. A 25 nm thermal oxide was grown to passivate the device before metallization. Different contacts were tested: Ti/Al and Ti/$TaSi_2$. The contacts were annealed at 600 °C in a forming gas. The Ti/$TaSi_2$ contacts performed better at high temperature. The amplifier demonstrated a bandwidth of approximately 10 kHz up to 450 °C. However, transimpedance gain increased with temperature. This was expected because sheet resistance increases with temperature. This early work, however, paves the way for future work in high-temperature sensor interface electronics. Alternate designs, including differential designs, may yield better overall consistency with temperature. This remains an open area of research.

Recently, operation at 500 °C of 6H-SiC JFET electronics for up to 7000 hours have been reported (Figure 3.16). These devices show a high degree of stability, with g_m variation less than 10 % and V_T only 1 % [35]. However, the same testing indicates other failure modes may be of concern. Although individual JFET, digital logic elements, and epitaxial resistors remained operational over prolonged exposure to high temperature, circuits of JFET devices stopped functioning at much shorter times. This suggests that the metal interconnects are more susceptible to failure even though the JFET structure is rugged. Examination points to the metal-dielectric interface as the weak point in these fabricated devices, presumably due to CTE mismatch between the materials.

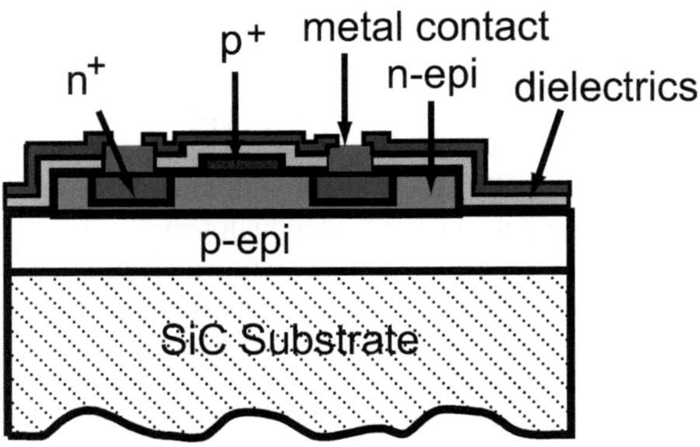

Fig. 3.15 Schematic cross-section of a n-channel JFET.

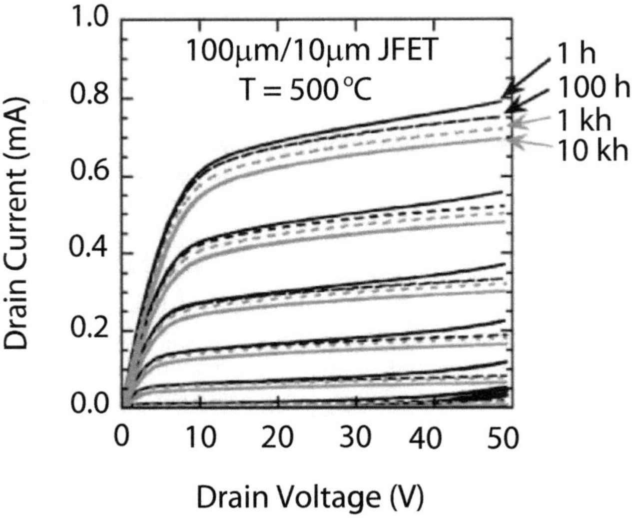

Fig. 3.16 [35] Drain I-V characteristics of packaged 100 μm/10 μm NASA 6H-SiC JFET measured during the 1st, 100th, $1{,}000^{th}$, and $10{,}000^{th}$ hour of electrical operation at 500 °C. Gate voltage steps are 2 V starting from $V_G = 0$ as the topmost sweep of each curve (©2009 Wiley-VCH), reprinted with permission.

3.5.3 BJT

The BJT topology is another interesting case for prolonged high temperature operation. Like the JFET, it utilizes a p-n junction, avoiding the need for high-quality dielectrics and buried metal features. A BJT device is expected to offer better on-state performance at higher temperatures than an equivalent JFET design, overcoming increased switching losses compared to a JFET [39]. Difficulty producing repeatable characteristics and lower stability in a minority-carrier device at operation above 300 °C have favored JFET designs over BJT [40]. For instance, BJTs require high quality metal contacts to both n- and p-type SiC regions, while JFETs are tolerant of non-ideal contacts. To date, there has been a lack of research using BJT high-temperature electronics because of these issues; however, operation at moderate temperatures have been proven out. For instance BJT-based transistor-transistor logic (TTL) built in 4H-SiC were successfully operated at 300 °C with little shift in logic characteristics [41]. Development of high quality contact metal stacks will open up further research using BJT circuitry for high temperature operation.

3.6 Electronics for communication electronics

Besides high temperature operation of power regulating circuits and sensor amplifiers, the harsh environment microsystem needs communication electronics. As has been discussed earlier, whenever possible, wireless is preferred. Assuming on-board power is an option (see Section 6.2), relying on wires for signal transmission adds extra levels of packaging and development of high temperature wire shielding and connectors.

Wireless electronics is typically divided into digital circuits for low frequency and analog circuits for high frequency operation. Silicon switching is currently used for up to 1 MHz operation and can be used to drive up to 1 MW, useful enough television broadcasting. SiC switches are being explored for these high power broadcast applications because both heat dissipation is enhanced in SiC components and SiC can withstand higher temperature operation. Together, this potentially reduces the need for expensive cooling systems used in commercial silicon switching installations.

Even when dealing with modest power transmission using SiC circuitry, significant issues need to be overcome. Deep levels are created in a wide bandgap semiconductor from metals or other defects incorporated into the material, which contain much more energy and have discharge times orders of magnitude longer than in silicon. Additionally, when being used for high power circuits, larger voltages are applied across the SiC structure. These two conditions together lead to a large population of electrons getting trapped in these deep levels, leading to unstable behavior at radio frequencies [42]. Hence, even higher purity base materials are needed compared to silicon. Furthermore, to date the successful power SiC topologies tend to rely on minority carrier injection, which slows down overall switching performance. MOSFET and MESFET technologies will be preferable for high frequency operation once the issue of poor dielectric layer formation is solved. Trying to borrow from successful silicon designs, heteroepitaxial configurations are also of interest. These topologies seem most promising by combining multiple SiC polytypes. However, this requires the substrate to be made from 3C-SiC for the junction to perform as desired, and high quality 3C-SiC substrates are a challenge to produce currently. Therefore, available topologies do not favor high frequency operation. Substrate cost is also a limitation for commercial applications such as cell phones.

Although harsh environment applications are less restricted in terms of overall device cost, harsh environment operation presents additional challenges for communication electronics. From a corrosion perspective, these electronics need to rely on contact metals that are not prone to oxidation or erosion. Tungsten, molybdenum, chromium, aluminum, nickel, titanium, gold, and platinum have all been explored as contact metals, but many of these readily oxidize. Creating a metal-oxide or metal-carbide capping layer can help protect the metal, but often these oxides and carbides degrade at high temperature. Hence, titanium, gold, and platinum are favored from a corrosion perspective. However, these metals can be difficult to deposit with low stress, which may cause mechanical failure under high temperature cycling. Additionally, high temperature operation may cause reduced transmission efficiency because resistivity increases and mobility decreases with increased temperature. Furthermore, discrete inductors are often used in communication circuitry. These devices would also need to be made to withstand high temperature or a highly corrosive environment as well.

These issues should be looked at as opportunities for further research. Solutions may well exist. However, research into the specific combination of high temperature and wireless power transmission has not been fully explored to date. For instance, further work in optimizing metal stacks for not only electrical contact stability but also general mechanical adhesion should lead to the desired reliability in metal interconnects. Dielectric development will improve the reliability of faster switching speed device topologies. Additionally, high temperature designs using low frequency transmission may be the preferable solution for harsh environment microsystems. SiC power devices have made significant development in the relatively short time that high quality substrates and epitaxial layer growth has been available. So it is not unreasonable to assume a similar pace can happen for high temperature SiC circuitry for sensor interface and signal transmission as well.

References

1. Feng ZC, Zhao JH (2004). Silicon carbide materials, processing and devices. Taylor and Francis, Inc. London UK
2. Rao, MV, Tucker, JB, Ridgway MC, Holland OW, Papanicolaou N, Mittereder J (1999). Ion-implantation in bulk semi-insulating 4H-SiC. Journal of Applied Physics 86(2):752–758
3. Rao MV, Tucker J, Holland OW, Papanicolaou N, Chi PH, Kretchmer JW, Ghezzo (1999). Donor Ion-Implantation Doping into SiC. Journal of Electronic Materials 28 (3):334–340
4. Gardner J, Edward A, Rao MV, Papanicolaou N, Kelner G, Holland OW (1997). Ion-Implantation Doping of Silicon Carbide. ORNL/CP-95077
5. Negoro Y, Kimoto T, Matsunami H (2003). High-Voltage 4H-SiC pn Diodes Fabricated by p-Type Ion Implantation. Electronics and Communications in Japan, Part 2, 86(12):434–441
6. Neudeck PG (2006). Silicon Carbide Technology. The VLSI Handbook, Chapter 5 (Editor Wai-Kai Chen, CRC Press, Second Edition)
7. Saddow SE, Williams J, Isaacs-Smith T, Capano MA, Cooper Jr. JA, Mazzola MS, Hsieh AJ, Casady JB (2000). Temperature Implant Activation in 4H and 6H-SiC in a Silane Ambient to Reduce Step Bunching. Materials Science Forum 338-342:901–904

8. Troffer T, Schadt M, Frank T, Itoh H, Pensl G, Heindl J, Strunk HP, Maier M (1997). Doping of SiC by Implantation of Boron and Aluminum. Physica Status Solidi A 162 (1):277–291
9. Kimoto T, Itoh A, Inoue N, Takemura O, Yamamoto T, Nakajima T, Matsunami H (1998). Conductivity Control of SiC by In-Situ Doping and Ion Implantation. Materials Science Forum 264–268:675-680
10. Choyke WJ, Matsunami H, Pensl G (2004). Silicon Carbide: Recent Major Advances. Springer-Verlag, Berlin, Heidelberg, New York.
11. Porter LM, Davis RF (1995). A critical review of ohmic and rectifying contacts for silicon carbide. Materials Science and Engineering B 34:83-105
12. Crofton J, L. M. Porter LM, Williams JR (1997). The Physics of Ohmic Contacts to SiC. physica status solidi B 202:581-603
13. Neudeck PG (2006). Silicon Carbide Technology. The VLSI Handbook, Chapter 5 (Editor Wai-Kai Chen, CRC Press, Second Edition)
14. Liu F, Hsia B, Senesky DG, Carraro C, Pisano AP, Maboudian R (2010). Ohmic Contact with Enhanced Stability to Polycrystalline Silicon Carbide via Carbon Interfacial Layer. Solid-State Sensor, Actuator and Microsystems Workshop, Hilton Head Island, South Carolina, June 6-10:214–217
15. Chung G-S, Yoon K-Y (2008). Ohmic contacts to single-crystalline 3C-SiC films for extreme-environment MEMS applications. Microelectronics Journal 39:1408–1412
16. Okojie RS, Lukco D, Chen YL, Spry DJ (2002). Reliability assessment of Ti/$TaSi_2$/Pt ohmic contacts on SiC after 1000 h at 600 oC. Journal of Applied Physics 91:6553–6559
17. Neudeck PG, Garverick SL, Spry DJ, Chen L-Y, Beheim GM, Krasowsk MJ, Mehregany M (2009). Extreme temperature 6H-SiC JFET integrated circuit technology. Physica. Status Solidi A:1–17
18. Sozza A, Dua C, Kerlain A, Brylinski C, Zanoni E (2004). Long-term reliability of Ti-Pt-Au metallization system for Schottky contact and first-level metallization on SiC MESFET. Microelectronics Reliability 44:1109–1113
19. Oder TN, Martin P, Adedeji AV, T. Isaacs-Smith T, William JR (2007). Improved Schottky Contacts on n-Type 4H-SiC Using ZrB2 Deposited at High Temperatures. Journal of Electronics Materials 36 (7):805–811
20. Teraji T, Hara S, Okushi H, Kajimura K (1997). Ideal Ohmic contact to n-type 6H-SiC by reduction of Schottky barrier height. Applied Physics Lettetter 71 (5):689–691
21. Lundberg N, Ostling M, Zetterling C-M, Tagtsrom P, Jansson U (2000). CVD-Based Tungsten Carbide Schottky Contacts to 6H-SiC for Very High-Temperature Operation. Journal of Electronic Materials 29(3):372–375
22. Oder TN, Sutphin E, Kummari R (2009). Ideal SiC Schottky barrier diodes fabricated using refractory metal borides. Journal of Vacuum Science and Technology B 27 (4): 1865–1869
23. Baliga BJ (2009). Advanced Power Rectifier Concepts. Springer Science+Business Media, LLC. New York, NY
24. Schoen KJ, Henning JP, Woodall JM, Cooper Jr. JA, Melloch MR (1998). A Dual Metal Trench Schottky Pinch-Retcifer in 4H-SiC. IEEE Electron Device Letters 19:97–99
25. Baliga BJ (1996). Physics of Power Semiconductor Devices. JWS Publishing.
26. Chow TP, Ramungul N, Fedison J, Tang Y (2004). SiC Bipolar Transistors and Thyristors. Silicon Carbide: Recent Major Advances, Choyke WJ *et al.* editors. Springer-Verlag, Berlin:737–764
27. Takayama D, Sugawara Y, Hayashi T, Singh R, Palmour J, Ryu S, Asano K (2001). Static and dynamic characteristics of 4–6 kV 4H-SiC SIAFETs. Proceedings of the 13th Inter. Symp. on Power Semiconductor Devices and ICs: 41–44
28. Palmour JW, Edmond JA, Kong HS, Carter Jr. CH (1993). Proc. of the 28th Inter. Society Engergy Converversion Conference: 1249–
29. Tan J, Cooper Jr. JA, Melloch MR. IEEE Electron Device Letters 9:487–
30. Ryu S-H, Agarwal A, Richmond J, Palmour J, Saks N, Williams J (2002). IEEE Electron Device Letters 23:321–
31. Agarwall A, Ryu S-H, Palmour J (2004). Chapter from Silicon Carbide: Recent Major Advances, Choyke WJ *et al.*, editors. Springer-Verlag, Berlin.

32. Neudeck P, Okojie R, Chen L (2002). High-temperature electronics–A role for wide bandgap semiconductors? Proc. IEEE 90(6):1065–1076
33. Neudeck PG, Spry DJ, Chen L-Y, Beheim GM, Okojie RS, Chang CW, Meredith RD, Ferrier TL, Evans LJ, Krasowski MJ, Prokop NF (2008). Stable Electrical Operation of 6H-SiC JFETs and ICs for Thousands of Hours at 500°C. Proc. IEEE 29(5):456–459
34. Patel AC (2009). Silicon Carbide JFET Integrated Circuit Technology for High-Temperature Sensors. Ph.D Dissertation, Electrical Engineering and Computer Science, Case Western Reserve University.
35. Neudeck PG, Spry DJ, Chen L-Y, Chang CW, Beheim GM, Okojie RS, Evans LJ, Meredith RD, Ferrier TL, Krasowski MJ, Prokop NF (2009). Prolonged 500 °C Operation of 6H-SiC JFET Integrated Circuitry. Materials Science Forum 615-617:929–932
36. Rebello NS, Shoucair FS, Palmour JW (1996). 6H silicon carbide MOSFET modelling for high temperature analogue integrated circuits (25-500°C). IEE Proc.-Circuits Devices Syst. 143(2):115–122
37. Brown DM, Downey E, Ghezzo M, Kretchmer J, Krishnamurthy V, Hennessy W, Michon G (1997). Silicon Carbide MOSFET Integrated Circuit Technology. Phys. Stat. Sol. A 162:459–479
38. Spry D, Neudeck P, Okojie R, Chen L-Y, Beheim G, Meredith R, Mueller W, Ferrier T (2004). Electrical Operation of 6H-SiC MESFET at 500 °C for 500 Hours in Air Ambient. Proceedings of IMAPS, Santa Fe, NM:WA1-1–WA1-7
39. Franke W-T, Fuchs FW (2009). Comparison of switching and conducting performance of SiC-JFET and SiC-BJT with a state of the art IGBT. 13th European Conference on Power Electronics and Applications:1–10
40. Neudeck PG, Garverick SL, Spry DJ, Chen L-Y, Beheim GM, Krasowski MJ, Mehregany M (2009). Extreme temperature 6H-SiC JFET integrated circuit technology. Phys. Status Solidi A:1–17
41. Lee J-Y, Singh S, Cooper JA (2008). Demonstration and Characterization of Bipolar Monolithic Integrated Circuits in 4H-SiC. IEEE Trans. Electron Devices 55(8):1946–1953
42. Morvan E, Kerlain A, Dua C, Brylinski C (2004). Development of SiC Devices for Microwave and RF Power Amplifiers. Silicon Carbide: Recent Major Advances, Choyke WJ *et al.* editors. Springer-Verlag, Berlin:839–867

Chapter 4
SiC MEMS devices

Silicon carbide microelectromechanical device development is currently a very active area of research with a primary focus on increasing the robustness of traditional silicon MEMS. Silicon MEMS has developed to a level of maturity in which several commercial ventures are deploying silicon technology into automotive and consumer electronics markets. A large knowledge base has developed into surface- and bulk-micromachined sensor types. This design expertise is being directly applied to sensor technology for harsh environment applications using SiC. This is accomplished because etch mask materials and etching techniques have already been developed that can be tailored to produce selective etching of SiC. These techniques can be linked together in a very similar fashion to silicon microfabrication, whether it be for surface- or bulk-micromachining of MEMS structures. This allows rapid development of SiC devices because the manufacturing concepts and design methodology can be readily applied. Hence, a large number of SiC MEMS sensor types have already been explored.

This chapter will first provide an overview of bulk- and surface-micromachining processes using SiC. This will be followed with a review of common MEMS sensor types as well as detailing research into using SiC to enhance traditional sensors. This will include a brief discussion of sensor design from the traditional application standpoint as well as specific considerations for deploying these sensor types into various harsh environment applications.

4.1 SiC MEMS processing

A generic SiC MEMS process flow is similar, although generally with fewer steps, than a generic IC process flow. Depending on the integration method, the MEMS process flow may start directly after the IC process flow or may start with the bare substrate. The primary differences between a MEMS process flow and IC process flow is the use of sacrificial etch layers to create free-standing structures and the use of deposited layers that are typically very thick compared to IC fabrication layers.

As will be detailed more fully in Section 4.2, the free-standing structures are used to create mechanical devices sensitive to various measurand (*e.g.*, strain, acceleration, pressure). The thick layers used in to create these devices are needed to increase the sensor sensitivity by increasing the sensing surface or increasing the mass of the free-standing structure. The following sections schematically outline the process flow to create a generic SiC MEMS device starting with a bare substrate using two general classes of fabrication, bulk- and surface-micromachining.

4.1.1 Surface (thin-film) micromachining

Surface-micromachining is basically an extension of the IC fabrication process (Section 3.1.1). Alternating insulative and conductive layers are deposited to create mechanically connected electrical routing. The main difference is the inclusion of a sacrificial release layer that can be etched away without damaging the layers above or below, enabling creation of free-standing structures. The surface micromachining process begins by depositing a non-sacrificial insulating layer to allow selective electrical isolation of the subsequent electrical routing layer (Figure 4.1a). This isolation layer can be silicon dioxide although layers such as silicon nitride or aluminum nitride are preferred because of their higher degree of chemical inertness compared to silicon dioxide as well as closer match in thermal expansion rate to SiC, which is important for high temperature applications. If electrical contact to the substrate is needed, a lithography and etch step is used to open select vias through this non-sacrificial insulating layer prior to deposition of the subsequent routing layer (Figure 4.1b).

Next an electrically conductive layer is deposited (Figure 4.1c). This layer is typically a thinner version of the later device layer to ensure compatibility with that deposition step, which in this case would be LPCVD 3C-SiC. The conductive layer is likewise patterned to define the electrical routing layer needed for device functionality (Figure 4.1d). This etch process needs to have a high selectivity to the non-sacrificial insulating layer (rapid etch rate of the target layer but very low to zero etch rate to other exposed layers). As was discussed in detail in Section 2.5, SiC etch techniques are more commonly plasma-based. Hence, the very high selectivity achieved for silicon using wet chemical etching is not available for SiC. However, silicon carbide etching processes can be made reasonably selective to silicon dioxide [1] and aluminum nitride [2]. Silicon nitride is less favorable in this regard, although it has been used successfully [3].

Next a sacrificial layer is deposited (Figure 4.2e). Conductivity is not important; however, the film must be able to be deposited relatively thick (typically one to a few microns). Most importantly, the layer must have an etch chemistry that has very high selectivity to the device and routing layer as well as to the non-sacrificial insulating layer. Borrowing from traditional MEMS processing, silicon dioxide is typically the sacrificial layer of choice with vapor or liquid hydrofluoric acid as the preferred release etch chemistry. However, because of high chemical resistance of

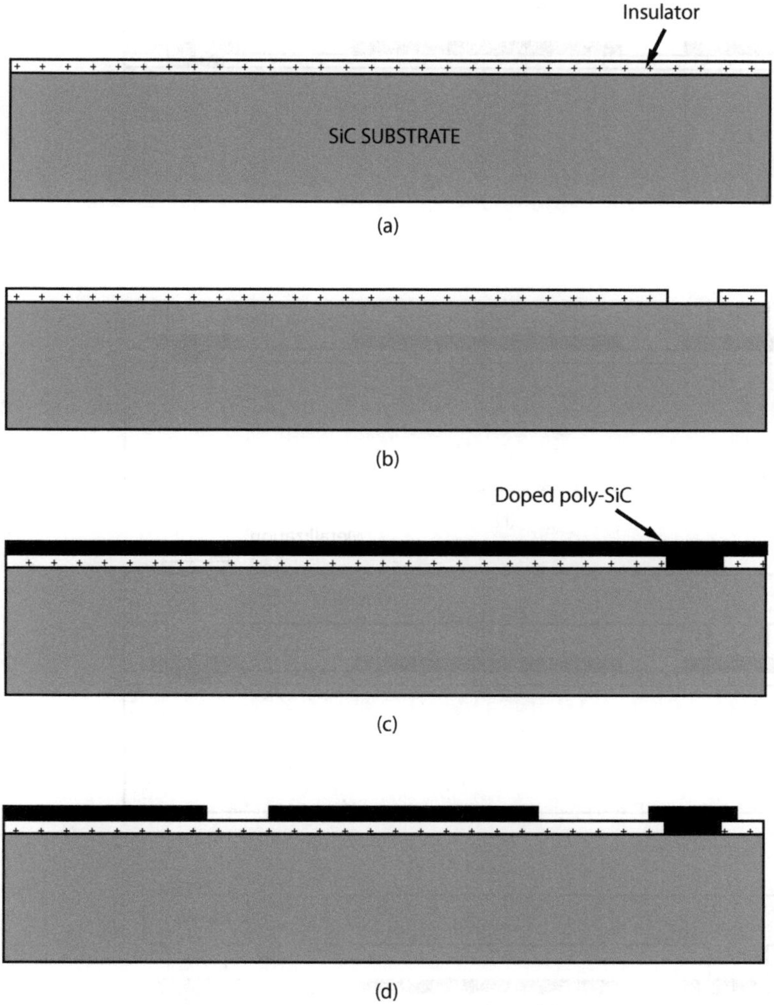

Fig. 4.1 Schematic of the first stages of fabrication of a single device layer surface-micromachined cantilever device with single electrical routing layer.

SiC compared to Si, a polysilicon layer is also a good option as a sacrificial material for SiC devices [6]. The sacrificial layer is typically patterned and etched to create via openings to the routing layer film (Figure 4.1f).

The sacrificial layer is followed with deposition of the SiC device structural layer (Figure 4.2g), which is again lithographically patterned and etched to define the device structure. The structural layer typically needs to be conductive, low-stress, and have a small stress gradient. This allows the layer to be shaped in a controlled way and maintain the desired gap dimensions around the structural device after the structure is released. Depending on its use, it may be important for the device

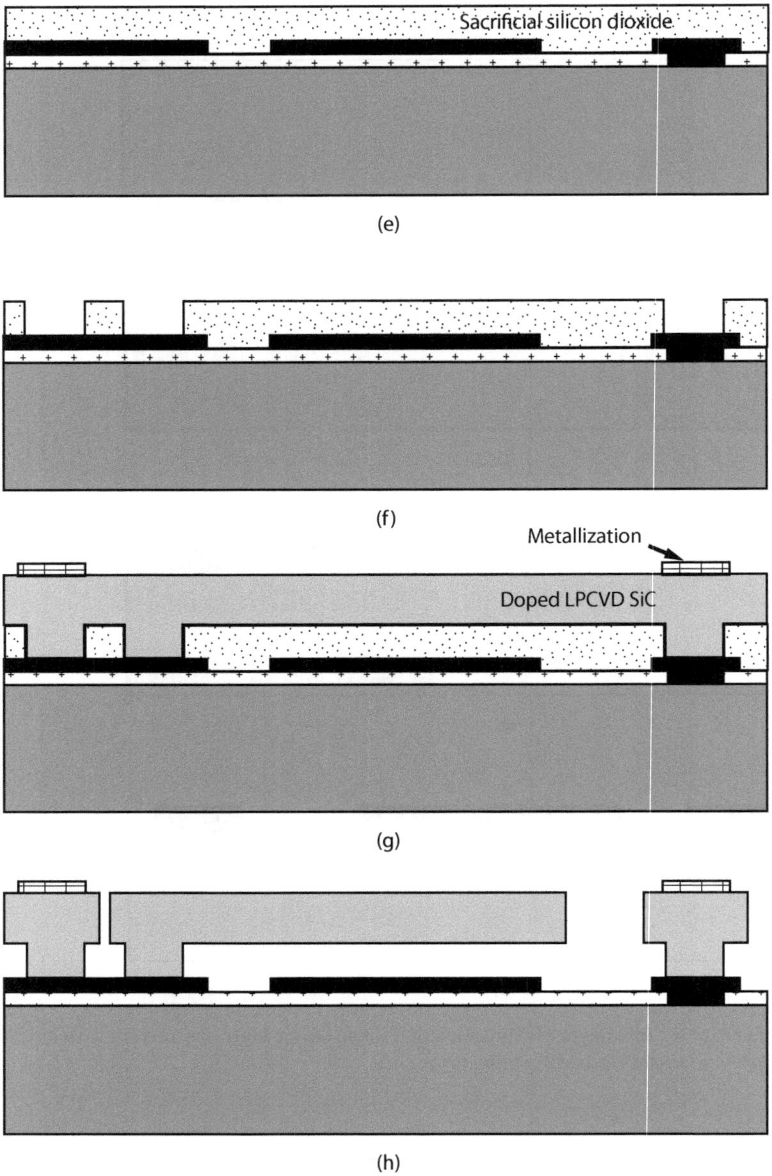

Fig. 4.2 Schematic of the final stages of fabrication of a single device layer surface-micromachined cantilever device with single electrical routing layer.

layer to also have proper ductility and fatigue characteristics. High stiffness allows for higher frequency operation. High yield strength prevents premature failure under shock or high acceleration events. The thickness of the device layer influences the resulting mass of the structure which is important for inertial sensing applications. Furthermore, in-plane sensing techniques will have signal amplitude that scales along with thickness, so the device layer thickness also influences the signal strength. Thickness also influences the stress developed during high-g acceleration. So proper design must consider all of these factors.

If needed for ohmic contact to wirebonds, a metallization layer will be deposited, lithographically patterned, and etched. Commonly the metallization is etched prior to patterning and etching of the device layer so that subsequent photoresist spinning does not need to coat and refill the deep trenches needed to penetrate the thick device layer. However, for simple structures in which the routing layer is omitted, the metallization can be deposited as a blanket film after the device layer is etched, using the device layer pattern to provide the necessary discontinuity in the metallization [4]. As was discussed in Chapter 3, Ti, Ni, and alloys with these metals are typically employed for this metallization layer, although other metals may be used depending on what corrosives the device will be exposed to during use.

Finally, the sacrificial layer is removed using a highly selective etchant (Figure 4.2h). Wirebonds are added to provide electrical connection to external power source and electronics if these are not integrated into the substrate.

Although it is not typically done, additional iterations of the insulating and routing layer deposition and patterning with additional interlayer insulator layers can be used to increase routing options depending on the electrical needs of a given sensor. Likewise, additional sacrificial and device layer pairs can be included in the overall MEMS fabrication flow to allow forming more complex structures including creating hinge structures or rotating gears. Fairly complex process flows have been successfully sustained for silicon MEMS fabrication [5]; however, this added complexity typically reduces yield or adds considerable cost. Therefore, many device design innovations in MEMS work to minimize the number of layers needed to fabricate the final structure.

4.1.2 Bulk micromachining

As was discussed in the previous section, device layer thickness can be a critical parameter in device design. Bulk-micromachining, where the substrate material is through-etched either to use the substrate as the device layer or a partial substrate etch is used to create a free-standing thin-film membrane, has become an equally important fabrication methodology in the manufacture of MEMS devices.

Because the substrate formation process is generally a high-temperature process not constrained to be compatible with other material layers, the resulting substrate fabrication process is optimized to create a high quality crystal with few defects and can be made with low stress. Additionally, the substrate is inherently thick compared to deposited films since it is created by slicing segments from a large boule of crystalline material (Section 2.2). Hence, substrates are generally hundreds of microns thick. Since selective etching processes have been developed for a wide range of substrate materials, including SiC, that are not high temperature, a bulk-

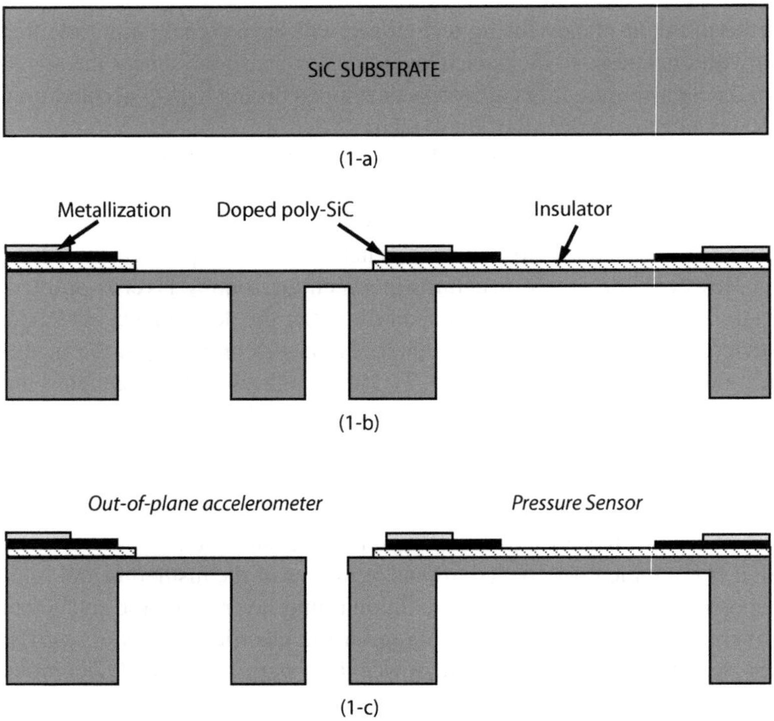

Fig. 4.3 Schematic of process flow to create bulk-micromachined cantilever and membrane devices using the substrate as a structural element. Additional detail is included to outline the basic process for incorporating piezoresistor elements into the structures for motion sensing.

micromachining process is attractive from a circuit integration perspective as well. (Integration issues will be discussed more fully in Chapter 6).

Figure 4.3 illustrates a generic process flow for bulk micromachining of a SiC substrate to create a free-standing beam and membrane. The process is comparatively simple and can be as simple as a single lithographic pattern and deep etch of the substrate; however, as is shown in the schematic flow, additional steps might be necessary to introduce other elements needed for sensing purposes. Etch characteristics, including selectivity to mask material and ability to etch high-aspect ratio trench structures, limits the usable substrate thickness and achievable gap dimensions, which can diminish some of the electrical signal gains achieved by having a tall structure. To achieve the necessary etch selectivity for bulk-micromachining, metal masking materials are typically required. Exposing the etch chamber to metal necessitates more frequent tool cleaning because the etch chemistry so slowly removes metal particles sputtered onto the chamber walls. Often fabrication facilities dedicate an etch tool to SiC etching using metal masks to prevent process cross-contamination. Furthermore, by etching completely through the substrate, the device now needs some encapsulation method to protect both sides of the substrate faces. (Encapsulation will be discussed in detail in Chapter 5).

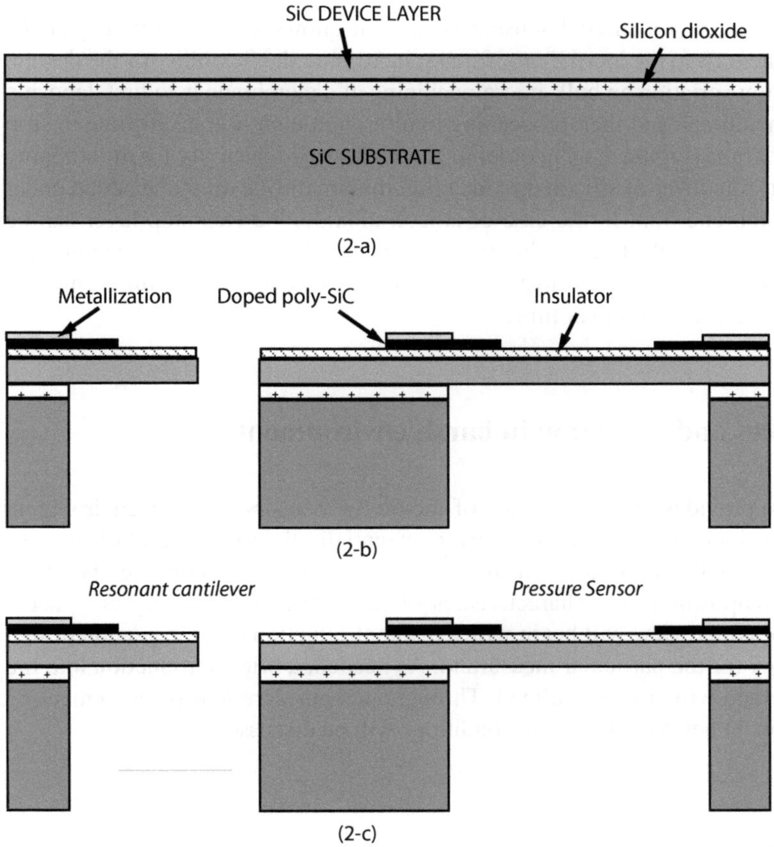

Fig. 4.4 Schematic of process flow to create bulk-micromachined devices by depositing a SiC thin-film layer and then etching the substrate up to the thin-film, which acts as the structural element. Additional detail is included to outline the basic process for incorporating piezoresistor elements into the structures for motion sensing.

Figure 4.3 illustrates a generic bulk-micromachining flow that utilizes a partial etch to create a free-standing membrane structure. The number of process steps is very small. In the case of the membrane (right side of Figure 4.3), additional structures will need to be fabricated on top of the membrane (such as thin-film piezoresistors) or onto a second wafer that can be be placed in close proximity to the membrane (such as a capacitor electrode) in order to detect membrane deflection. In order to sense pressure, the membrane must be subjected to a pressure differential, which necessitates isolating the frontside of the membrane from the back. This can be accomplished by embedding the substrate into another component to provide the necessary isolation or by creating a reference cavity behind the membrane by bonding another substrate below the etch pit. By continuing with the partial trench process and then patterning a through-etch region (Figure 4.3-1-b and -1-c, right side), it is possible to make a free-standing cantilever with a substantial end-mass.

This can be useful for inertial sensing because it allows some decoupling of the spring stiffness from the inertial proof-mass, increasing the flexibility of the design.

A common variation on bulk-micromachining of a membrane is to first deposit a thin-film membrane and then proceeding to a through-etch of the substrate to stop on the membrane (Figure 4.4). In order to get the desired selectivity for the stopping layer, a thin-film layer of silicon dioxide or aluminum nitride might be added under the SiC membrane film. In the case of silicon dioxide, the etch stop layer can be easily etched away selectively. Although this may not be necessary, silicon dioxide is not as robust a mechanical material so may be removed to minimize shifts in membrane characteristics over time.

4.2 Devices and operation in harsh environment

This section provides a review of state-of-the-art SiC sensors for harsh environment sensing. Harsh environment sensing covers essentially the same gamut of measurands as traditional commercial sensing: pressure, acceleration, strain, temperature, chemical composition, and characteristics of fluids such as viscosity or opacity. Each sensor type review will begin with a brief introduction into the general design methodology for the particular measurand. As needed, a brief introduction into the sensing method will also be outlined. Throughout, considerations of implementing these sensors in harsh environment conditions will be discussed.

4.2.1 Strain

Various structures — ship and spacecraft hulls, aircraft wings, pressure vessels, industrial piping — experience variable structural loads due to weight, expanding gases, or variations in temperature. For these type of structures, real-time monitoring of structural deformation either by monitoring peak dynamic load or by long-term monitoring of minute drift in shape (creep) can be used to signal scheduling preventative maintenance, remove the structure from active operation, or to restrict operation to a reduced-load operating condition (e.g., reducing process pressure to prevent structural failure). These structural changes can be monitored through the use of strain sensors mounted on the surface of the structure.

Strain is a useful metric of deformation and creep of a mechanical structure. The one dimensional measurement of engineering strain when the underlying strain field is constant or linearly varying with position is given by:

$$\varepsilon = \frac{\Delta l}{l_0} \tag{4.1}$$

This change in dimension can be measured purely as a change in length. Strain measurement techniques that measure the displacement over a fixed length (referred to as the gauge length) that results from strain in a structure are referred to as extensometers. Extensometers are an indirect measurement of strain. This was originally a common method of measuring strain in macro-scale applications. However, these have been largely displaced by thin-film piezoresistive strain gauges (metal-foil gauges).

Some materials experience changes in electrical properties due to strain in the material. This effect is called a "piezo" response. Two common classes are piezoresistive and piezoelectric. Piezoelectric response is exhibited by certain materials such as quartz and aluminum nitride when the material is in a highly oriented crystalline form. Piezoelectric response of SiC films is very poor, so this mechanism will not be discussed in detail here, although the interested reader is directed to [7] for more information.

Most semiconductor materials exhibit a large piezoresistive response, including polycrystalline SiC [8]. Piezoresistivity is a change in electrical resistance with applied strain. The degree of piezoresistivity can be tailored somewhat by the type and amount of doping in the semiconductor. When strain is applied to a piezoresistor, the electrical resistance of the film changes because of the three-dimensional change in the structure due to Poisson coupling as well as an inherent change in resistivity due to the applied strain. Taking the example of a slender piezoresistive trace deposited directly onto a substrate that can experience mechanical deformation, the change in resistance is given by:

$$\frac{\Delta R}{R_0} = (1+2\nu)\varepsilon + \frac{\pi}{\xi}\varepsilon \tag{4.2}$$

where ξ is the resistivity of the film, ν is Poisson's ratio, π/ξ is the piezoresistance constant, and the subscript 0 refers to the unstrained state. Generally, it is desirable to have L be much larger than the width and strain gauge thickness should be minimized to ensure a uniform strain field through the entire thickness. Minimizing the thickness also prevents locally stiffening the structure, which can alias the resulting strain measurement [9]. The piezoresistance constant (gauge factor) for metal resistors such as Ni are on the order of 1 to 2. For semiconductors such as Si or SiC, it can be between 50 and 150 [8].

Typically, the measurement monitors the resulting voltage changes across the resistor while applying a known, constant current. It is certainly possible to use a frequency-based metric such as pairing the resistor with a capacitor to form a RC-tank, which has a frequency that is a function of resistance [10]. Because of electrical noise and thermal variations, it is also common to use multiple resistive elements in a Wheatstone bridge configuration to negate these variations [11].

Because of the very high accuracy with which changes in capacitance and frequency can be measured, strain using micromachined transducers is also measured using these techniques. These are extensometer devices. They may attempt to very accurately measure minute displacements or may involve mechanical structures for amplifying motion due to strain to increase the output displacement.

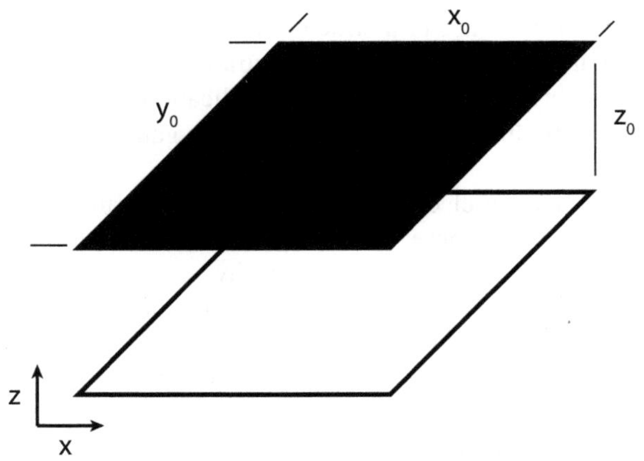

Fig. 4.5 Schematic of two square capacitor plates.

The capacitance between two parallel conductive plates (Figure 4.5) is given by the following equation:

$$C = \frac{\varepsilon x_0 y_0}{z_0} \tag{4.3}$$

where ε is the permittivity of the dielectric and is often written as some multiple of the permittivity of free space, ε_0 ($\varepsilon = \varepsilon_r \times \varepsilon_0$). The overlapping area of the plates is given by the lateral dimensions x_0 and y_0 while the distance between the plates is given as z_0. To allow freedom of motion, the dielectric is typically air or partial vacuum, thus $\varepsilon \approx \varepsilon_0$.

From this equation, it is seen that the capacitance can be varied mechanically by either changing the amount of overlapping area of the two plates:

$$\frac{\partial C}{\partial x} = \frac{\varepsilon y_0}{z_0} \tag{4.4}$$

or by changing the spacing between the plates:

$$\frac{\partial C}{\partial z} = -\frac{\varepsilon x_0 y_0}{z_0^2}. \tag{4.5}$$

For micromachined strain sensors, the intent is typically to measure either in-plane film stresses or to measure external strain applied to the substrate. In these cases, the capacitor plates are formed between beams cut out from the device layer. Example configurations are shown in Figure 4.6.

Changing the spacing between the plates (referred to as gap-closing mode, Figure 4.6b) is typically ten to one hundred times more sensitive than changing the amount of overlap area for the same initial gap dimension. In both modes, the change in capacitance is increased, hence the resolution is increased, if the initial gap dimension can be decreased or the surface area can be increased. The latter is accomplished by

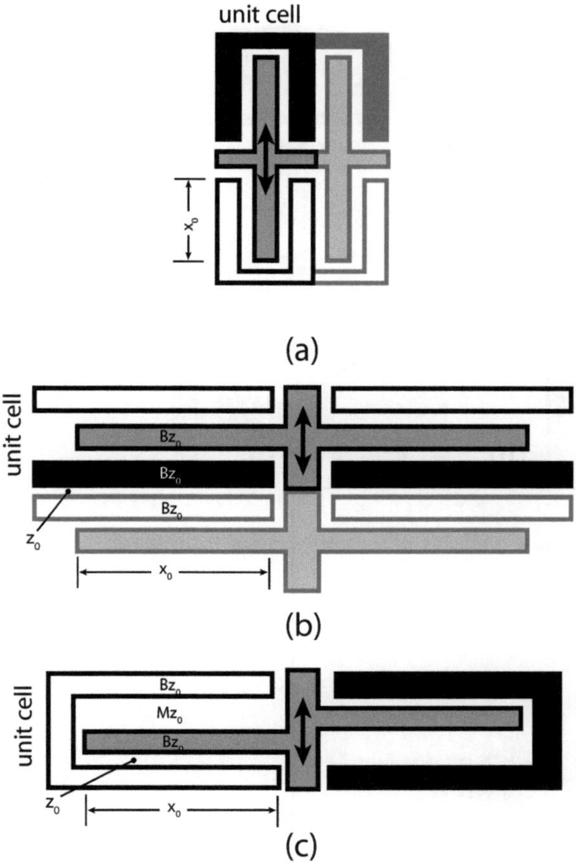

Fig. 4.6 Schematic of (a) lateral motion, interdigitated finger capacitor arrays and (b) gap-closing capacitor array utilizing independent plates on each side of the moving plate and (c) an offset-gap array when the stationary plates are not fully electrically isolated from each other.

increasing the device layer thickness and electrically connecting multiple plate pairs in parallel (plate arrays). The former though suffers from various limitations. In the gap closing mode, the linear region of response is limited to displacements less than about one tenth of the initial gap. This can severely limit the full-range strain measurable by the gauge. Both modes are limited by practical limitations in etching deep gaps, compromising the usable film thickness, which decreases the capacitor plate area. Also, very small gaps are more prone to contact each other either due to strain overload, external vibration (shock), or by the attractive force that forms in the capacitor when charge is applied.

For methods that use the change in overlap area (lateral mode), the most common topology is an array of interdigitated fingers, referred to as comb arrays (Figure 4.6a). For strain sensing, a chevron [4] or other lever mechanism is used to transform the small distortion due to strain into larger motion of the comb. Although the

(a) (b) (c)

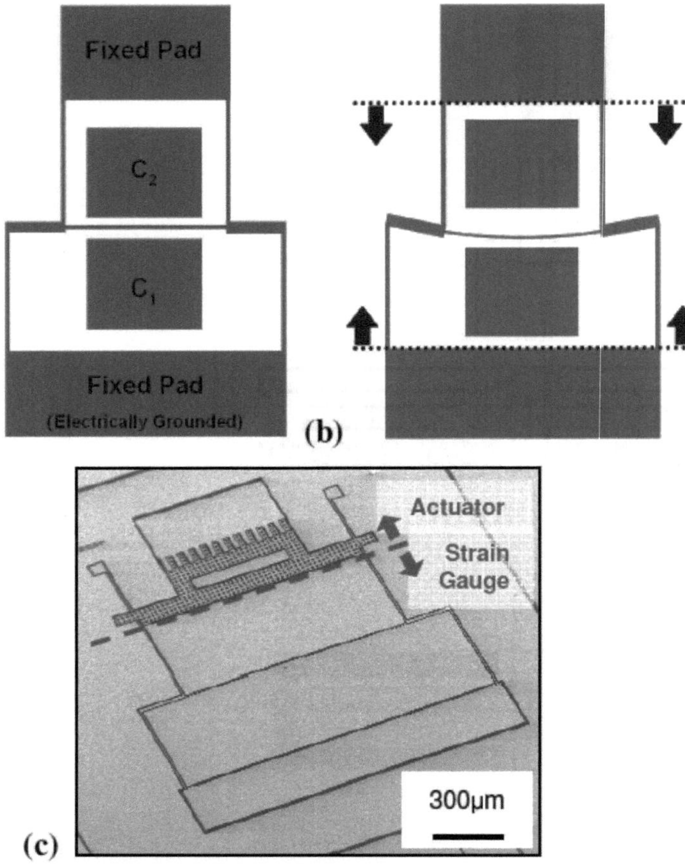

Fig. 4.7 [12] Schematic of four-point bending strain gauge design and SEM image of fabricated device coated with LPCVD SiC (©IEEE 2007), reprinted with permission.

lateral mode is linear over large displacements, the lever mechanism needs to be precise since any errors will also be amplified. Likewise, these amplifying mechanisms increase sensitivity to off-axis and temperature-induced strain. So all these factors need to be taken into account into the final design. Alternatively, mechanisms can be used to minimize degradation of on-axis signal while decreasing the influence of off-axis sensitivity, as was done by Jamshidi *et al.* [12]. In this case, an in-plane, four-point bending plate is loaded using slender tethers (Figure 4.7). The bending plate is part of a gap-closing capacitor, which is used to measure applied strain. The slender tethers along with having the bending plate longer than the fixed plate makes the structure insensitive to off-axis strain. This structure has been demonstrated as a SiC coated Si device fabricated from a SOI wafer and tested up to 370 °C.

Alternately, a resonant structure can be anchored between two points such that the anchors define the gauge length. Such a configuration will create a tension in the resonant structure as a result of strain applied between the anchor points. Although any doubly-anchored structure can be used for this type of sensing, clamped-clamped beam and double-ended tuning fork (DETF) resonators are most common. Their resonant frequencies exhibit a high sensitivity to tension in the beam while being relatively insensitive to off-axis strain because of the large difference in structural compliance along the beam compared to perpendicular to the beam.

For a clamped-clamped beam with an attached proof mass, its resonant frequency of first bending mode can be written as follows:

$$f = \frac{1}{2\pi}\sqrt{\frac{198EI}{0.198\rho AL + kM_c} + \frac{4.85E\varepsilon}{0.198\rho AL + kM_c}} \tag{4.6}$$

where E is the Young's modulus of the beam, ρ is the density of the beam, I is the area moment of inertia, A represents the cross-sectional area of the beam, L is the beam length, M_c is the static mass of any structure attached to the beam such as a comb array, and k is a scale factor for this attached mass depending on location of attachment and the particular mode shape being excited during resonance. (For most practical applications the attached mass is at the center of the beam and the fundamental mode is used, in which case $k = 1$.) For details on how the constants are found for this mode shape, the interested reader is directed to [13] for details on theory of vibration. Typically MEMS resonators of this type range in frequency from 10s to 100s of kHz.

From Equation 4.6, one can calculate the change in frequency with applied strain. Normally, this result is non-dimensionalized by dividing the change in frequency by the nominal resonant frequency of interest:

$$\frac{1}{f_0}\frac{\partial f}{\partial \varepsilon} = \frac{2.425}{4.85\varepsilon_0 + 198I} \tag{4.7}$$

which, if the initial strain is zero, further reduces to

$$\frac{1}{f_0}\frac{\partial f}{\partial \varepsilon} = \frac{0.01225}{I} \tag{4.8}$$

Equation 4.7 can be used with ε_0 not equal to zero to account for residual stress in the device layer film by converting this stress into an equivalent strain using Hooke's Law. This implies that high residual tensile stress in the device layer film reduces the sensitivity of the resonator frequency to strain, which is consistent with experimental evidence.

Although this provides a means for strain to induce a change in frequency, the minute motion of the resonator needs to be detected in order to measure frequency. This is typically accomplished using gap-closing capacitive plates or lateral motion comb arrays. These are used not only because of their high sensitivity to motion, but also the capacitor can also be used to induce an in-plane force on the beam to excite and maintain the resonator vibration. From Equation 4.7, it is also clear that adding this mass comes at a cost of reducing the frequency to applied strain. So again the overall design must weigh the benefits of the various topologies into the overall system requirements.

Initial work in harsh environment resonant strain sensing took conventional epitaxial Si strain sensors and coated them with a 30 nm thick LPCVD SiC coating [14]. These sensors were operated above 180 °C as well as were subjected to a potassium hydroxide (KOH) wet etch solution heated to 80 °C for 5 minutes. The coating was shown to protect the underlying Si. However, very high temperature operation (above 400-500 °C) requires converting this design to SiC due to Si softening.

A DETF strain sensor from polycrystalline 3C-SiC device layer has also been developed. Besides a change in device layer material, the work by Azevedo, *et al.* aimed to increase device robustness to high g-shock events [15]. A variation on standard DETF topologies was implemented to maintain the high transduction area of designs using comb sense arrays while preventing this suspended mass from causing failure due to stiction or excessive stresses in the structure. This is accomplished by relocating half of the comb drive structure between the tines to balance the mass symmetrically around the tines (Figure 4.8). Utilizing a square-wave drive oscillator topology [16], this 3C-SiC DETF resonant strain sensor achieved a comparable strain resolution to the standard comb-driven DETF topology built from an epitaxially-grown Si device layer (0.045 $\mu\varepsilon$ in a 10 kHz bandwidth, Figure 4.9). This 3C-SiC strain sensor has been successfully operated up to 600 $°C$ in steam and subjected to 64,000 g shock without damage to the structure [3].

4.2.2 Pressure

Another common harsh environment sensor is a pressure sensor. SiC or SiC-coated Si pressure sensors are expected to find their greatest utility in monitoring gas or liquid pressure in industrial chemical applications. In addition, they will also find specialized use for space mission and medical applications.

Pressure sensing is typically accomplished by either monitoring membrane deflection under a pressure differential or by how pressure variation impacts the resonant frequency of a vibrating structure inside the variable pressure region. Membrane deflection is by far the most common pressure sensing method.

Membranes can be either round or rectangular. Maximum deflection occurs at the center of a symmetric membrane. For small deflections (less than half the membrane thickness), center-point deflection of a flat membrane under a static pressure differential is given by [17]:

$$\Delta z = \frac{pa^4}{64D} \tag{4.9}$$

for a circular membrane and

$$\Delta z = \frac{pa^4}{50D} \tag{4.10}$$

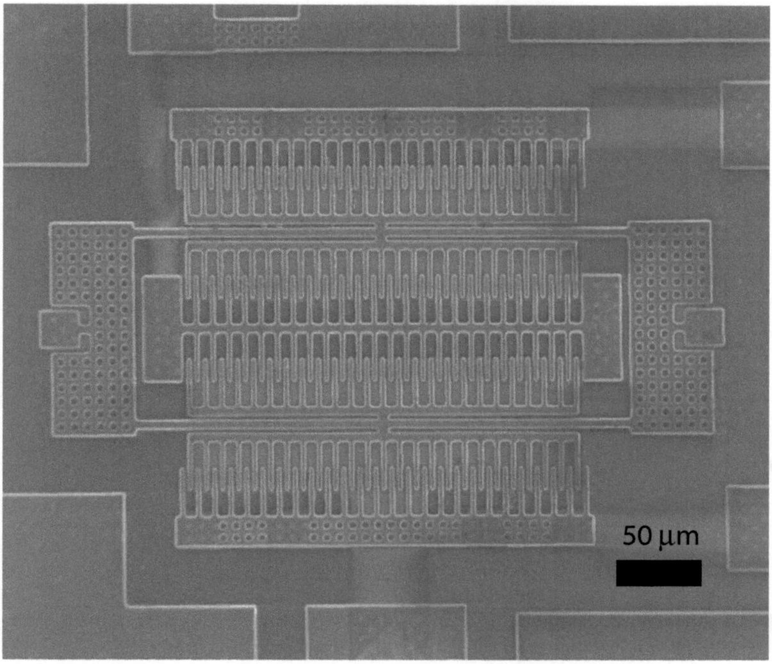

Fig. 4.8 [15] SEM image of a balanced double-ended tuning fork (BDETF) resonant strain sensor fabricated in a 4 mask process using LPCVD polycrystalline 3C-SiC as the device layer (©IEEE 2007), reprinted with permission.

for a square membrane, where a is either the radius or the half-length of the side of the square, p is the uniformly distributed pressure differential. D is the flexural rigidity and is related to the elastic modulus and plate dimensions:

$$D = \frac{Eh^3}{12(1-\nu^2)} \tag{4.11}$$

where h is the plate thickness and ν is Poisson's ratio.

As the deflection increases beyond half the membrane thickness, a cubic term response becomes significant due to the influence of shear stress in the membrane [18]. This can limit the full-range pressure measurable while maintaining output voltage linearity. In order to extend the linear response pressure range, a bossed membrane can also be used (Figure 4.10). For a circular bossed membrane, small deflections of the center boss as a function of pressure is given by [19]:

$$\Delta z = \frac{pa^4}{64D}\left(1 - \frac{b^4}{a^4} - 4\frac{b^2}{a^2}\log\frac{a}{b}\right) \tag{4.12}$$

where b is the radius of the boss feature. Although the boss stiffens the membrane

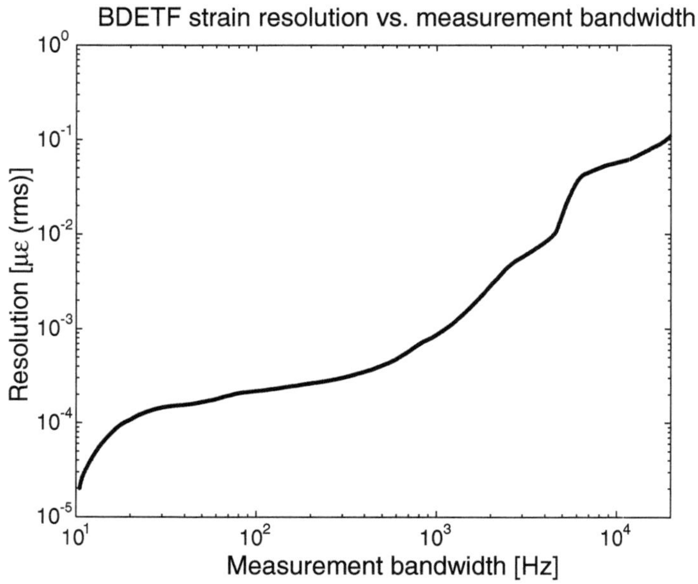

Fig. 4.9 [15] Experimental measurement of strain resolution as a function of measurement bandwidth for the 3C-SiC BDETF strain sensor ((©IEEE 2007), reprinted with permission.

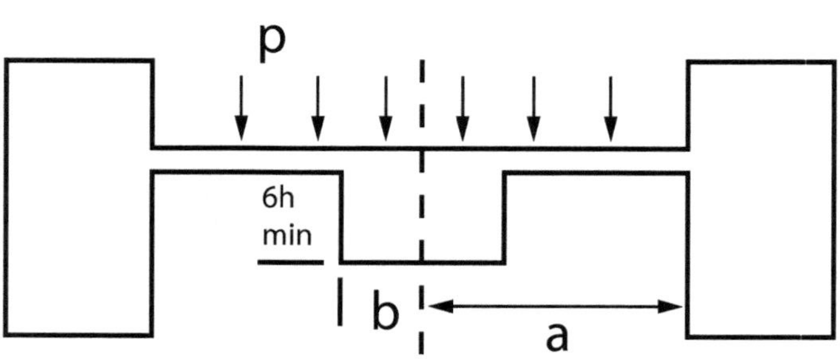

Fig. 4.10 Schematic of a bossed membrane diaphragm.

for a given size, it increases the linear response range for a given membrane size as well.

Small deflections of the membrane require sensitive measurement techniques. Semiconductor-based piezoresistive measurement techniques [11] convert strain in the resistor region into a change in electrical resistance, which can be used in a voltage sensing circuit to determine membrane deflection. Typically, a Wheatstone bridge configuration is used with the piezoresistors positioned so that strain increases resistance in one pair of resistors and decreases in another pair of resistors by positioning the resistor elements in tensile and compressive regions of the membrane, respectively. Peak radial strain occur at the perimeter of the membrane and can be related to pressure. From [18], this relationship for a flat membrane is:

$$\varepsilon_r = -\frac{3}{64}\frac{a^2 h}{D} p \tag{4.13}$$

Developing this sort of relationship allows depositing thin-film piezoresistors onto a diaphragm in order to accurately measure the pressure differential across the membrane.

Early pressure sensor work utilized doped 6H- or 3C-SiC as the piezoresistive element, although the diaphragm was either created by bulk-etching Si under the piezoresistive elements [21] or utilizing an SOI wafer to create a Si diaphragm. In this latter case, the 3C-SiC layer used to create the membrane could be selectively deposited on Si using an LPCVD process with methylsilane as the precursor gas and a fast carbonization step using ethane [22]. Deposition was carried out at 1150 °C. The Si device layer was thinned down to 200nm in part by thermally growing a 1400 nm thick silicon dioxide insulating layer over the device layer. This design utilized a bossed membrane and exhibited a 35mV/V sensitivity at 200 °C.

Current state-of-the-art SiC pressure sensors include depositing NiCr piezoresistive elements on a silicon dioxide layer to electrically isolate them from a n-type 3C-SiC film, which was grown on a Si substrate. Bulk chemical etching of the Si is used to form a free-standing SiC membrane [20]. A sensor constructed this way has been tested up to 400 °C with a full-scale pressure range of 500 kPa. It exhibits a temperature sensitivity of -0.16%/°C at 400 °C.

In order to make a pressure sensor capable of withstanding even higher temperatures and be less sensitive to corrosive environments that might etch NiCr, an all-SiC structure has also been developed. It was formed by first forming p-type 6H-SiC piezoresistive elements on a n-type 6H-SiC substrate. The substrate is then bulk etched using a photoelectric-enhanced chemical etch to form a 50 μm thick membrane [21]. This pressure sensor has been experimentally shown to exhibit stable pressure response up to 500 °C with a full-scale pressure range of 1000 psi. It exhibits a gauge factor temperature sensitivity of -0.11%/°C at 500 °C.

An alternate sensing technique is to use the membrane as one side of a gap-closing capacitor. In the case of a capacitive readout pressure sensor, typically the capacitor is formed by using the deformable membrane as one of the plates and placing it in relatively close proximity to a fixed electrode plate. A circular bossed membrane pressure sensor is well modeled using equation 4.14. In this case, a change in the gap distance, z, result in a change of capacitance. As a function of pressure, change in capacitance around the zero-deflection point is given by:

$$\frac{\partial C}{\partial p} = \frac{\pi p b^2 a^4}{64 D z_0^2}\left(1 - \frac{b^4}{a^4} - 4\frac{b^2}{a^2}\log\frac{a}{b}\right) \tag{4.14}$$

where the area is defined by the size of the boss, which is simply $A = \pi b^2$, where b is the radius of the boss. However, for a flat membrane pressure sensor, the actual change in z varies as a function of the radial position of the membrane. Hence, the actual change in capacitance calculation requires integrating the change in z with radial position. Using the standard deflection model for a clamped circular membrane under small deflection, one can determine a relationship between pressure

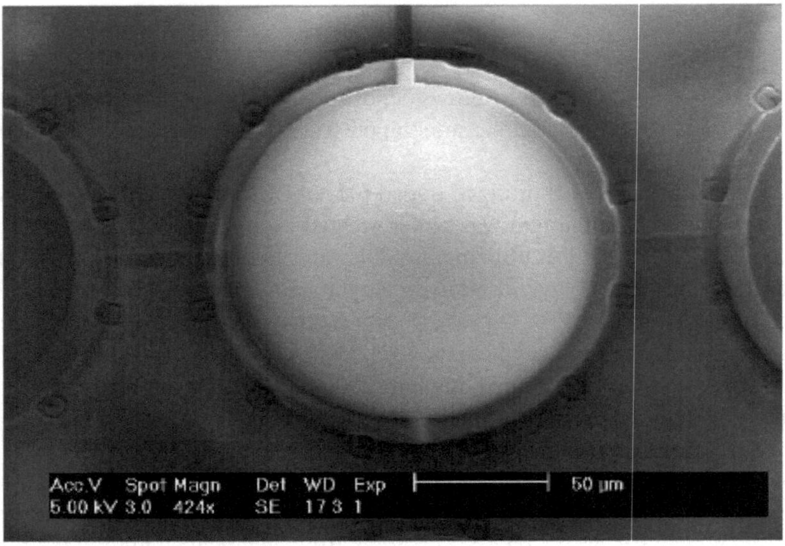

Fig. 4.11 [24] SEM image of a PECVD SiC pressure sensor (©IOP 2004), reprinted with permission.

and capacitance [23]:

$$C = \frac{\varepsilon a^2}{4h} \frac{(-ln(8\sqrt{hD} - \sqrt{p}) + ln(8\sqrt{hD} + \sqrt{p}))}{\sqrt{p}} \tag{4.15}$$

As is seen from equation 4.15, this relationship is in general non-linear, restricting operation to a relatively small pressure range if linear output is desired.

Surface-micromachined pressure sensors are of interest because the final sensor design is simpler since the backside of the diaphragm does not need to be encapsulated, and some bulk etching techniques are not compatible with integrated circuits. The mechanical quality of the sensor diaphragm is not as good as one formed from the substrate material because the substrate is a single-crystalline, very low-impurity material, whereas surface-micromachined diaphragms are typically a polycrystalline or amorphous film and may have non-uniform crystal growth or high residual stress. Process research aims to minimize these differences though because of the greater potential for integration with sense electronics and simplified packaging. An example of a surface-micromachined pressure sensor made from SiC was fabricated using PECVD a-SiC as the diaphragm material (Figure 4.11). Experimental measurements demonstrate a sensitivity of 3.2 pF/MPa and nonlinearity of 2% from 0.001 to 0.101 MPa for an array of 100 diaphragms that are 80 μm in diameter [24]. In this work, sputtered Al was used to form the fixed and moving electrode. The moving electrode is sandwiched between two SiC layers to prevent potential exposure of the Al to corrosive environments.

Because the deflection is restricted at the edges, the overall change in capacitance is significantly reduced without a boss structure. This in turn leads to the need for a very small gap, which can limit the ability of the pressure sensor to survive overpressure events. One way to overcome this is to introduce a dielectric coating over the stationary electrode, which prevents the two electrodes from shorting out when large pressures are applied. This technique has been successfully used for pressure sensing over a very wide range of pressures by purposely operating where the normal range of pressures causes the center of the diaphragm to contact the dielectric. After contact, increasing pressure leads to an increase in contact area. Although this leads to a number of nonlinear responses, experimental results show regions of decent linearity over a portion of the very wide range of pressures achievable with this type of sensor [25]. By having the critical gap dimension defined by a thin film dielectric instead of by etching means it can be manufactured with a highly controlled thickness down to nanometers. Furthermore, sensor output is increased because typically these dielectrics have a dielectric strength that is several to tens times greater than an air gap.

This touch-mode capacitive pressure sensor has been recently created out of an entirely SiC family of components [26]. The routing and diaphragm layers were formed from ammonia-doped 3C-SiC deposited using LPCVD (Figure 4.12). A Ni/Cr/Au stack was used as the high-temperature contacts. These sensors were tested up to 5 MPa (700 psi) static pressure. Although test chamber limitations prevented high-temperature testing at pressures above 100 psi (non-contact regime), a 172 μm diameter diaphragm demonstrated a sensitivity of 7.2 fF/psi at 574 °C. These sensors were successfully used for in-cylinder combustion pressure measurements on a test-bench setup.

Another method for pressure sensing in the vacuum regime is by monitoring the quality-factor (Q) of a resonator. In this case, by monitoring the frequency output of a resonator and watching either the 3dB width of the resonant signal amplitude peak or monitoring shifts in frequency due to Q variation, the pressure of the atmosphere in contact with the resonator can be determined. As the vacuum increases, the molecules around the resonator become less dense, leading to less collisions — the air damping decreases. Because air damping is an energy loss mechanism in a resonator, increased damping broadens out the resonant peak width in the frequency domain. Although this technique has been explored using silicon resonators, it has yet to be demonstrated using SiC.

The resonant structure for this method of pressure sensing can take a variety of forms, including the topologies laid out in Section 4.2.1. It can also take the form of a large diaphragm vibrating out of plane. Only, unlike in the previous diaphragm examples, this time a force generator is required to drive the diaphragm into the desired mode shape. This can be accomplished using a capacitive plate actuator or by depositing piezoelectric elements onto the membrane. Likewise, this technique has been applied to Si micro-sensors; however, it has yet to be demonstrated in SiC technology.

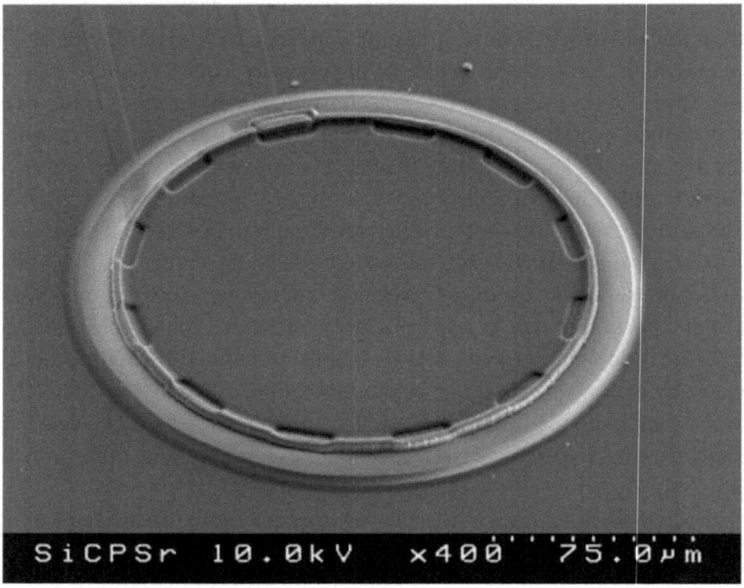

Fig. 4.12 [26] SEM image of LPCVD surface-micromachined pressure sensor element (©Elsevier 2008), reprinted with permission.

4.2.3 Inertial Sensing

Inertial sensing units are useful in motion sensing and navigation applications. Inertial sensing involves accelerometers, gyroscopes, and an accurate timing reference. Position information can be obtained by the double integration of acceleration with respect to time. So, if acceleration is known in three orthogonal directions along with a precise measurement of time, precise knowledge of position can be determined. Likewise, by knowing the rate of rotation about three orthogonal axes, precise orientation can be determined. By combining all of these components together, six degrees of freedom can be determined, creating a full description of a body's motion. Typical methods for determining acceleration and rate of rotation will be discussed along with specific examples of SiC versions intended for harsh-environment applications.

Accelerometers work by monitoring displacement of a well-characterized, movable mass (proof-mass) or by capturing the reaction force needed to hold the mass in place under an acceleration event. In the case of a displacement sensor topology, the mass is suspended such that the stiffness of the suspension is small in one direction but large in all other directions. Thus, the motion of the proof mass can be attributed to the component of acceleration that is aligned with the non-stiff direction of the suspension. In this case, resolution of the accelerometer is dependent on both the resolution of motion detection and the size of the proof mass since a larger mass will result in a larger deflection for the same suspension stiffness and acceleration.

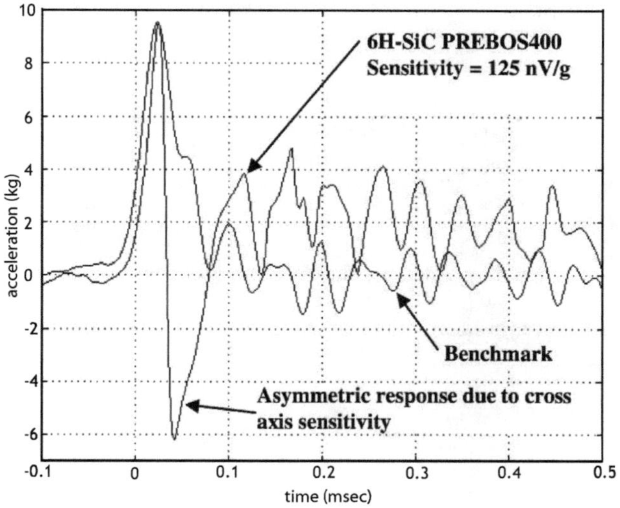

Fig. 4.13 [28] Comparison of out-of-plane membrane-based 6H-SiC accelerometer to a commercial high-g accelerometer (©Elsevier 2003), reprinted with permission.

Alternately, one can measure the reaction force needed to hold the proof mass in place. This force is directly proportional to the acceleration. A similar configuration for suspending the proof mass that limits proof mass motion to be preferential in one direction is often used so that the force sensor design can be optimized to measure a single-axis force. Force sensing accelerometers generally utilize the same phenomenon of varying tension in a vibrating element that is used in resonant strain sensors [27]. By including force actuators such as capacitive plates that can push back on the proof-mass until the detected displacement is reduced to zero, the voltage necessary to maintain a zero displacement condition becomes a measurement of the applied acceleration. By keeping the proof-mass stationary, nonlinearities are generally avoided and coupling to rotational acceleration during proof-mass motion is reduced.

SiC accelerometers to date have been of the displacement measurement type. Initial work used bossed membranes made from 6H-SiC substrates to detect out of plane acceleration [28]. Square and circular bossed diaphragms were fabricated and benchmarked against a commercial accelerometer (Endevco 7270-060 K). The sensors were well matched for accelerations below 8,000 g (Figure 4.13). However, above 40,000 g the SiC accelerometer showed non-linearities believed to be either due to residual stresses in the membrane or slight misalignment of the piezoresistive sense elements to the membrane. Although the boss increases the sensor sensitivity by increasing the proof mass, a flat membrane design was also developed because the boss tended to increase off-axis sensitivity as well by inducing a twisting motion.

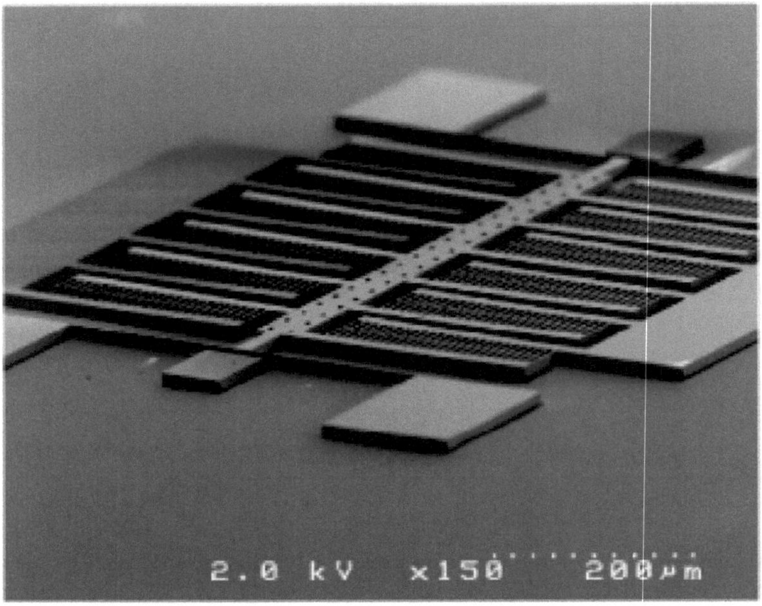

Fig. 4.14 [29] SEM image of an in-plane 3C-SiC accelerometer utilizing capacitive comb drives as the sense mechanism (©TTP 2009), reprinted with permission.

More recently, advances in strain gradient control and precision reactive ion etch micromachining have allowed development of in-plane single-axis SiC accelerometers. Utilizing large capacitive comb arrays as both part of the proof mass and the motion sensing mechanism, accelerometers created from LPCVD 3C-SiC have demonstrated a resolution of 350 $\mu g/\sqrt{Hz}$ at 1 kHz [29] (Figure 4.14).

PECVD a-SiC has also been used to fabricate in-plane and out-of-plane accelerometers [30]. Sensors made from a-SiC must also utilize a conductive layer for capacitive position readout. In this work, aluminum was used as the conductor (Figure 4.15). These sensors exhibit 1.8 fF/g and 2.3 fF/g sensitivities in the vertical and lateral directions, respectively. Because of the choice of materials and resulting low temperature deposition processes (> 400 °C), these sensors are potentially Si CMOS compatible; however, the low deposition temperature makes PECVD films susceptible to creep if used for higher temperature applications.

Timing references do not need to use a mechanical element. A simple RC tank can be used as a timing reference; however, such a resonant system usually has high energy loss, which makes the drive circuitry power inefficient. By introducing a mechanical element into the oscillator circuit, a high frequency, high Q (tight frequency band) signal can be generated and used to keep track of elapsed time. The higher the precision that time is measured, the more accurate the position can be determined from rate of rotation and acceleration information.

Folded flexure resonators have been made by directly growing polycrystalline 3C-SiC on silicon dioxide on a Si substrate. Aluminum is used as the etch mask

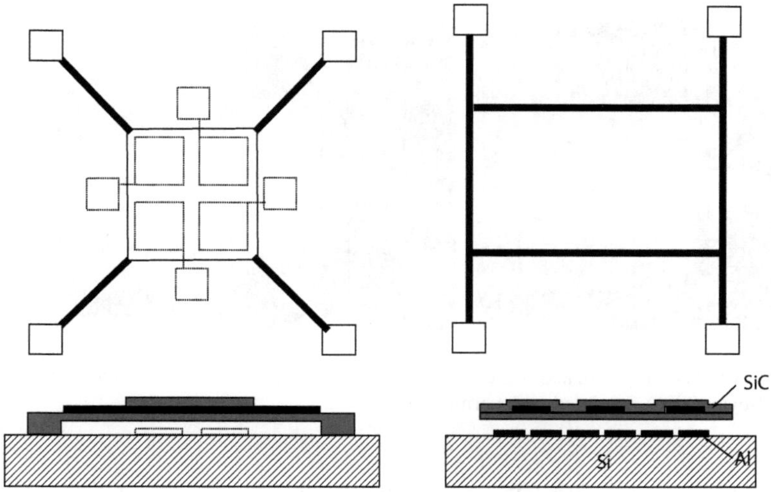

Fig. 4.15 [30] Schematic of layout for (left) out-of-plane and (right) in-plane accelerometers fabricated from PECVD a-SiC (©IEEE 2003), reprinted with permission.

material. This type of structure has an in-plane resonant frequency of 42.6 kHz with a 110 V actuation voltage [31]. These resonators were tested up to 900 °C. It was shown to exhibit significantly less shift in resonant frequency with temperature than an equivalent polysilicon resonator, presumably because of a higher sensitivity of Young's modulus to change in temperature for the polysilicon beams [32].

The DETF structure used as a strain sensor is also a potential timing reference. The resonant strain sensor operates at 218 kHz. However, it was only tested at atmospheric conditions, thus exhibiting a fairly poor Q of approximately 300-400 [14]. Even a simple cantilever structure can be used as a mechanical resonator. For instance a vertically-actuated SiC cantilever beam exhibited a resonant frequency of 1.5 MHz and a Q of 128 [33]. If these devices were operated under vacuum, significantly higher Q should be achievable.

A Lamé mode resonator array has also been demonstrated in polycrystalline 3C-SiC deposited using LPCVD. A fully-differential array of resonators exhibited a Q of 9,300 at a frequency of 173.5 MHz, operated in air [34]. The intended application was as a mechanical filter; however, it demonstrates higher frequency operation that could also be used for timing applications. A limitation of polycrystalline 3C-SiC resonators continues to be higher motional resistance. Even though very small capacitive gaps were formed using FIB in this particular example, motional resistance was still high for this resonator array. This is presumably because the sheet resistance in poly-SiC films is still significantly higher than other common materials, such as poly-Si. Nonetheless, as a basic resonator timing element, the Lamé mode resonator is a viable choice for SiC-based timing circuits.

Although the above resonators were all capacitively driven and sensed, thermally-actuated resonators have also been demonstrated (Figure 4.16). By patterning Pt or

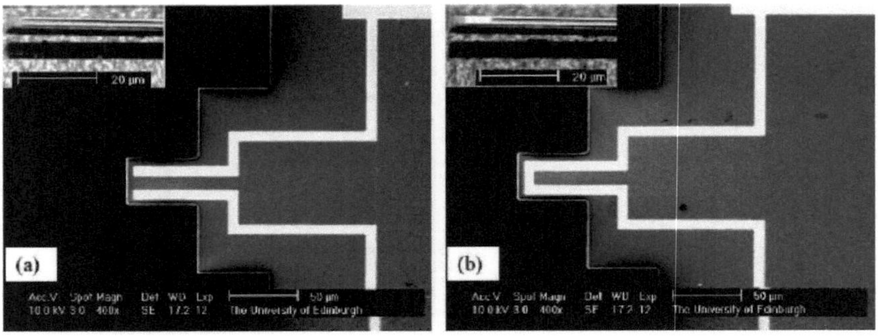

Fig. 4.16 [35] Two thermally-actuated SiC cantilever resonator designs. Pt is used as the electrode material (©Elsevier 2006), reprinted with permission.

Ni/Cr electrodes on hetroepitaxially-grown SiC cantilevers, resonant frequencies near 900 kHz have been achieved [35].

Gyroscopes are used to measure rate of rotation. This is accomplished most typically by a mass-in-frame configuration where a mass is vibrated in a direction orthogonal to both the axis of rotation of interest and the direction to which the frame is restricted to react. Monitoring either the position or reaction force generated by the frame due to a Coriolis acceleration that is generated during rotation gives a measurement of the rotation rate.

This type of single-axis gyroscope, using a variety of mechanical suspension and frame designs, has been implemented successfully in Si technology [36, 37, 38]. To date, no known micromachined gyroscope has been fabricated in SiC. With the successful implementation of strain sensors and accelerometers, SiC microfabrication has demonstrated all the necessary components to fabricate this type of device. Thus, it is just a matter of time before this last component for inertial sensing is realized in SiC.

Integrating the accelerometer and gyroscopes together in a fashion similar to newly emerging commercial Si-based inertial measurement units (IMUs) will be a challenge for the high-end temperature applications. The combined packaging of these devices will need to be developed using packaging ceramics that can survive the high temperature without generating large stresses in the package due to thermal expansion mismatch. Additionally, reduction of electrical interconnects is needed to improve the robustness of the packaged solution under these high temperature and potentially corrosive environments. These first-level packaging issues will be discussed more fully in Chapter 5. These complications need to be weighed against the process and design difficulty of requiring even higher degrees of monolithic integration of the IMU sub-components.

4.2.4 Chemical

There are a number of chemical sensors that utilize SiC films. Key advantages of SiC include chemical inertness and temperature stability without significant oxide growth or inter-metal diffusion. Being chemically inert allows the SiC films to be subjected to harsh chemical conditions without destroying the film. Minimal oxide growth and material property stability with temperature allows these sensors to maintain a higher level of stability than Si- or polymer-based sensors. Likewise, the low diffusion rate of metals into SiC maintain sensor stability when higher temperatures would otherwise suffer from metal migration into the ceramic films, changing their electrical characteristics.

Chemical sensors can be implemented using elements that are sensitive to adsorption of chemicals onto the surface. This includes capacitors, Schottky diodes, FETs, and surface acoustic wave (SAW) devices. In some instances SiC is not used directly as part of the sensing element, but may merely be a tough mechanical support that is tolerant to high localized temperature, particularly for use in micro-hotplate fabrication. In this case, the sense material is typically heated to stabilize the operating temperature or to improve sensitivity by promoting a chemical reaction.

Hydrogen is a common target chemical. By placing a catalytic metal (*e.g.*, Pt) onto a semiconductor surface, hydrogen atoms, either from a catalytic reduction of hydrogen molecules or hydrocarbons at the metal surface, diffuse through the metal to form an electrically-polarized layer at the metal-to-semiconductor interface [39]. When made part of a capacitor, this polarized layer lowers the flat band voltage. When part of a diode, it reduces the barrier height. By monitoring how the electronic device varies, the amount of hydrogen can be determined.

Oxygen atoms or nitric oxide molecules consume hydrogen, so tend to result in the reverse effect. Carbon monoxide reacts with oxygen, which in turn inhibits that oxygen from reacting with hydrogen. These competing mechanisms through their association with hydrogen can be sensed using hydrogen sensitive devices.

Two classes of SiC chemical sensors will be discussed. The first group are field effect devices, which use transistor of diode devices to detect the changes in electrical polarity due to hydrogen build-up. The second group do not use SiC to form the sensor elements but instead use it as a high-temperature, high-strength substrate for creating a suspended micro-hotplate to heat the sense elements.

4.2.4.1 Field Effect Devices

In 1997 metal-insulator-SiC (MISiC) sensor devices based on capacitors and Schottky diodes were tested [40]. These devices were constructed from 4H-SiC, and operating temperatures as high as 1000 °C were demonstrated with prolonged operation at 600 °C feasible. Response times were on the order of milliseconds. Additionally, interdiffusion of metal and insulator layers were studied. Response curves were generated for hydrogen, oxygen, and hydrocarbons.

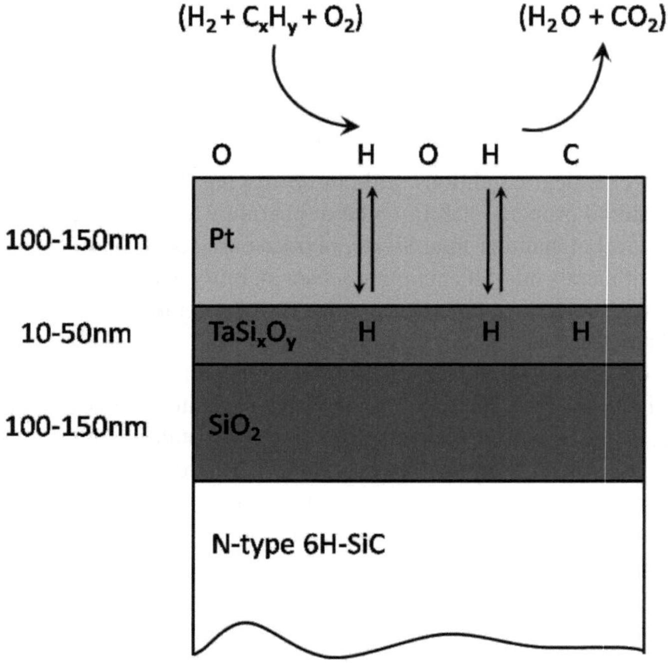

Fig. 4.17 [41] Schematic cross-section of a MISiC sensor that incorporates $TaSi_x$ as a buffer layer. This buffer layer interacts with the silicon dioxide layer to form a $TaSi_xO_y$ layer. (©Elsevier 1997), reprinted with permission.

Baranzahi *et al.* [41] further demonstrated 6H-SiC MISiC devices that incorporated a "buffer" layer between the metal and silicon dioxide (Figure 4.17). This buffer of 10 nm thick $TaSi_x$ improved the long term stability and response time of the sensors. These sensors were used to monitor hydrocarbons and were able to exhibit an output transition near a stoichiometric ratio in a simulated exhaust environment, which indicates complete combustion.

A MISiC structure that incorporated a LaF_3 buffer layer to increase the sensitivity to fluorine was developed in 1999 [42]. This structure was originally developed using a Si semiconductor layer. However, high temperature was needed for the catalyst and buffer layer to create the desired sensitivity. So the semiconductor layer was switched to SiC so that operation over 200 °C was feasible. This sensor was tested against F_2, HF, CCl_4, CH_4 and sensitivity to more complex fluorine compounds such as CF_3CH_2F to see if these compounds could be differentiated in the output. It was found that although there sensitivity was highest in the range of 200-330 °C, selectivity was not possible. But at 390 °C, selectivity was possible (Figure 4.18).

Work utilizing LaF_3 was then further extended to an array of MISiC sensors that utilized variations in metal catalyst (*e.g.*, porous Pt, Ir, Pt/Ir) to generate differentiating responses in a real exhaust environment. Figure 4.19 shows a schematic of the

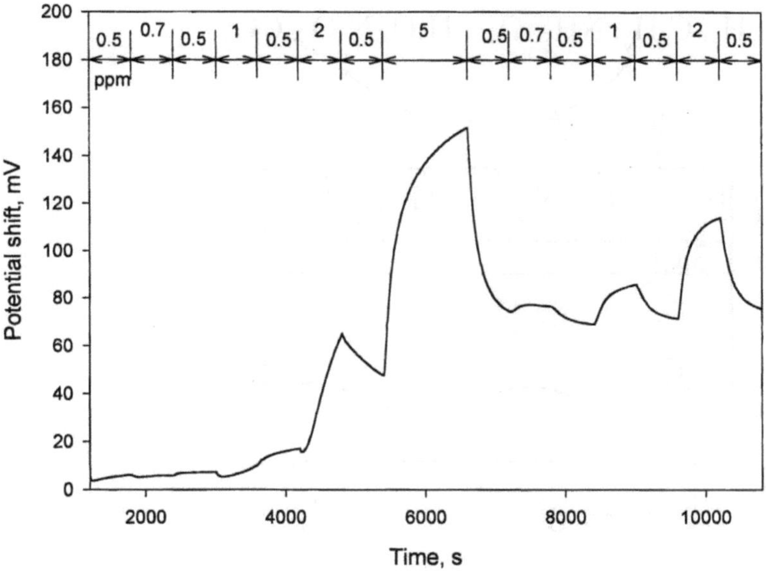

Fig. 4.18 [42] Response of a SiC gas sensor to HF at varied concentrations at 390 °C (concentration of HF in ppm is given in the upper portion of the graph). (©Elsevier 1999), reprinted with permission.

device utilizing a porous catalyst layer to increase the tri-interface regions as well as the response curves for such a device when exposed to an oxygen containing environment versus propane [43]. FETs became the device of choice because they combined the simplistic readout electronics of diode-based devices with the stability of capacitor-based devices since the FET devices also utilize a thicker oxide layer. This device is also easily adaptable to SiC, making it possible to operate these types of gas sensors at high temperatures either to improve sensitivity or because the application demands survival at elevated temperature, as is the case of in-cylinder combustion monitoring.

Although this array approach was shown to be somewhat successful, the general drawback of these types of sensors is the lack of a single sensor to a single chemical. So for practical implementation an array of sensors is likely needed to make the necessary distinction, which in turn requires a detection algorithm that can be quite complex depending on the models used [44].

4.2.4.2 Micro-hotplates

Micro-hotplates are useful for chemical sensing because control of the sense material temperature is advantageous in two main ways. First, actively regulating the sense element temperature to a constant level dramatically reduces thermally-induced drift or aliasing of the output signal. Second, in some cases by heating the

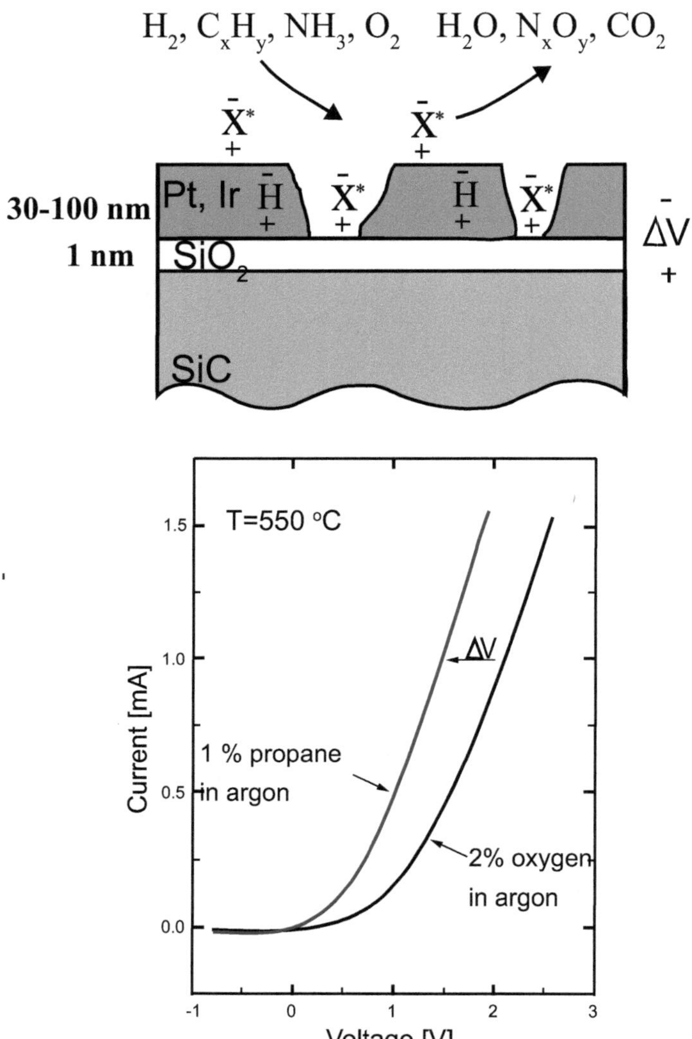

Fig. 4.19 [43] (a) Schematic of the MISiC sensor utilizing porous Pt and (b) response curve for this type of sensor when exposed to a propane or oxygen environment (©Elsevier 2001), reprinted with permission.

sense element to a high temperature leads to improved sensitivity to a given species or allows operating in a regime where response specificity among a selection of chemicals is greater.

Well-controlled, localized heating using resistive trace elements is typically achieved by creating a free-standing suspension on which the resistive elements are fabricated into and capped with metal oxides. SiC is a great candidate for these free-

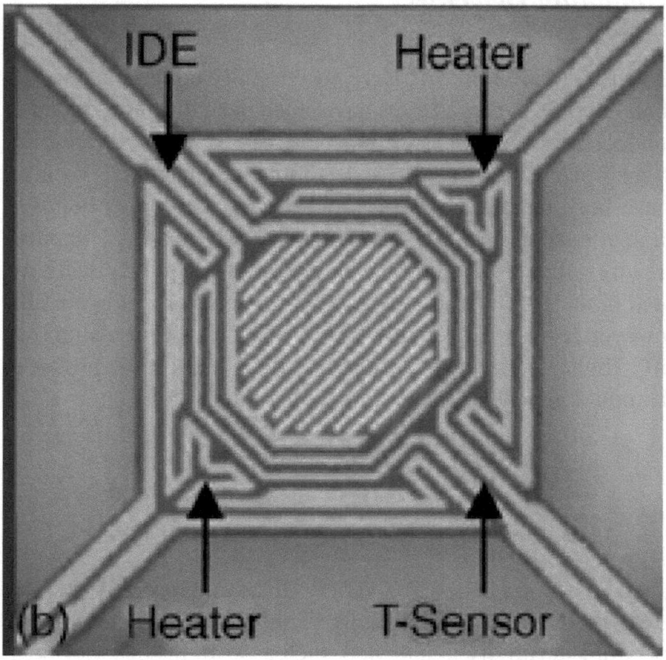

Fig. 4.20 [45] Microphotograph of micro-hotplate made on top of a 2 μm thick LPCVD 3C-SiC film (©Elsevier 2005), reprinted with permission.

standing suspensions because of the high mechanical strength, mechanical stability at high temperatures, and chemical resistance. Because silicon can be selectively etched with respect to SiC, these suspensions are typically deposited onto silicon substrates so that the silicon can simply be etched away under the SiC.

One example of such a design used a 2 μm thick LPCVD 3C-SiC film deposited on a Si substrate [45]. This film was covered with a 300 nm thick PECVD silicon oxide to isolate the catalyst from the substrate. Patterned Pt was used as the resistive heating element, temperature sensor, and inter-digitated electrode for chemical sensing. Finally, a second layer of PECVD oxide was deposited over the structure and gold was used to overcoat the bond pad regions (Figure 4.20). The micro-hotplate was able to heat up to approximately 500 °C using 300 mW, demonstrating the high temperature survivability of SiC. Continued work in this area should look at high ambient temperature chemical sensing. In this case, the micro-hotplate would likely need to be transitioned to a SiC substrate. It would also need to be operated at an even higher temperature than ambient to maintain a consistent chemical sensor temperature relative to changes in ambient.

4.2.5 Photo and Radiation Detectors

Light emitting diodes (LEDs) based on SiC were successfully commercialized in 1989 by Cree Semiconductor [46]. These LEDs have since been displaced by other wide-bandgap semiconductors because of low efficiencies of SiC-based devices. Reversing the light emitting process allows sensitive detection of light in the UV range. SiC photodetectors have been made on 4H- [47] and 6H-SiC substrates [48, 49]. These sensors have high efficiencies for UV wavelengths (~93% external quantum efficiency [47]) but are insensitive to visible wavelengths, allowing good selectivity.

SiC was investigated in the 1960s for use as a radiation detector, but again efficiencies were low; however, recently Stoken *et al.* have revisited SiC as a alpha ray and x-ray detector [50]. Higher purity levels of the SiC films that can be produced today have improved output efficiency compared to earlier attempts.

4.2.6 Temperature

Temperature sensing can be accomplished in a variety of ways. Generally, devices exhibit different degrees of temperature sensitivity. This is often something one wants to avoid. Temperature sensitivity in a device, such as a strain sensor or pressure sensor, can sometimes be counteracted by using an independent temperature sensor along with a model of the sensor response to temperature integrated into the sensor output electronics. Conversely though, isolating the strain sensor from strain essentially converts the sensor output to one that is solely a function of temperature. Therefore, optimizing temperature sensitivity instead of avoiding it can convert resistors, transistors, and resonators into usable temperature sensors.

Measuring resistance changes in a surface-micromachined thin film ceramic strip creates a low-complexity, compact temperature sensor. Termed thermistors, these temperature sensors have a long history. Many researchers have developed thermistors from SiC films and have studied the performance of the different polytypes with different dopant types and levels. In 1982, Nagai *et al.* demonstrated a SiC thermistor fabricated using magnetron sputtering onto an alumina plate [51]. These devices were trimmed to control the nominal resistance value prior to sealing inside a hermetic glass package (Figure 4.21). The resistance of these devices as a function of the inverse of temperature is shown in Figure 4.22. Thermistors have also been demonstrated using patterned metal electrodes on unintentionally doped 3C-SiC CVD wafers [52] as well as a high density array formed using magnetron sputtered SiC [53]. The array was specifically made into a linear array with regions as dense as 50 μm center-to-center spacing to characterize the thermal gradient at the liquid-vapor interface during evaporation (Figure 4.23).

In 1996, Casady *et al.* demonstrated a temperature sensor based on a 6H-SiC buried gate junction gate field-effect transistor [54] (Figure 4.24). The *pn* junction current was regulated using a series-series feedback network to prevent fluctuations. In this configuration, the *pn* junction voltage, v_d, was shown to have good sensitivity

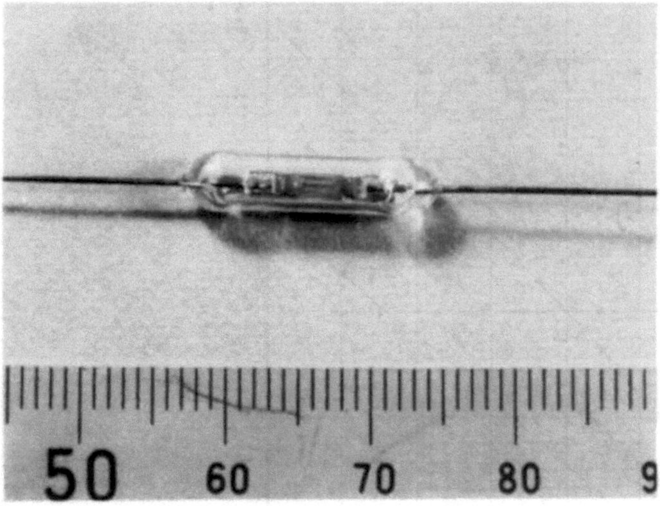

Fig. 4.21 [51] Hermetically sealed SiC thermistor (©IOP 1982), reprinted with permission.

(2.3 mV/°C) and excellent linearity from 25 °C to 500 °C. The sensitivity rapidly increased at -50 °C however, presumably because of carrier freeze-out in the diode.

Recently, a temperature sensor based on the optical signature of a SiC chip in a sintered SiC tube has been successfully used to accurately measure high temperatures up to 750 °C [56]. This sensor uses blackbody radiation, two-wavelength pyrometry along with Fabry-Perot laser interferometry to provide relatively fine temperature accuracy. Furthermore, this method uses the SiC chip as a free-space optic, which provides wireless operation. The use of a laser system is inherently bulky compared to a simple semiconductor resistor or JFET temperature sensor and requires, at least in the current configuration, that the laser be kept in a low temperature environment. This method though has promise in combustion engine or turbine environments where the high temperature environment is localized in the system.

4.3 Material aspect of harsh environment operation

This section will briefly discuss issues related to high temperature operation. Namely temperature-induced stresses will be discussed. Furthermore, control of stress gradient and its impact on being able to recreate certain devices that exist in Si technologies will also be discussed.

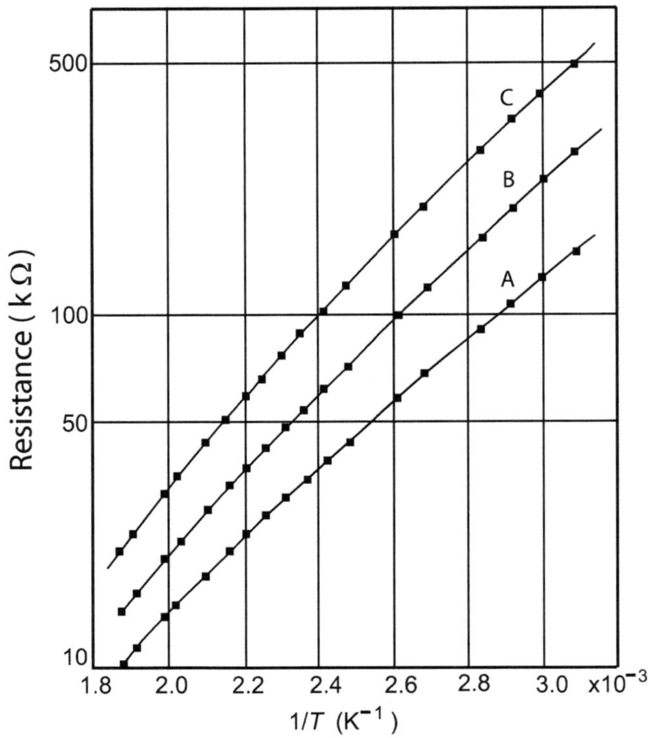

Fig. 4.22 [51] Temperature dependence of resistance of the SiC films deposited at various substrate temperatures with a particular electrode pattern: A 750 °C; B 700 °C; C 650 °C (©IOP 1982), reprinted with permission.

4.3.1 CTE of layers and effect

When operating temperatures can be hundreds of degrees Celsius over room temperature and dynamic temperature ranges can likewise be quite large, even small differences in how materials expand with increased temperature can be of critical importance. This is often captured as a coefficient of thermal expansion (CTE). CTE mismatch is a critical input in designing a sensor for harsh environment applications. Furthermore, CTE is generally not a linear function of temperature [57, 58, 59]. Figure 4.25 plots CTE values for SiC and Si based on a variety of deposition methods. It is interesting to note that the common CTE values listed in textbooks for Si and SiC do not apply to the same temperature range. In fact, near room temperature the CTE difference becomes very small.

In the case of a SiC double-anchored, flexural-beam resonator fabricated on silicon, the mismatch between the silicon substrate and the SiC device layer film leads to a higher frequency sensitivity to temperature than would be expected by elastic modulus softening of SiC alone (see Equation (1.13)). There also exists a region

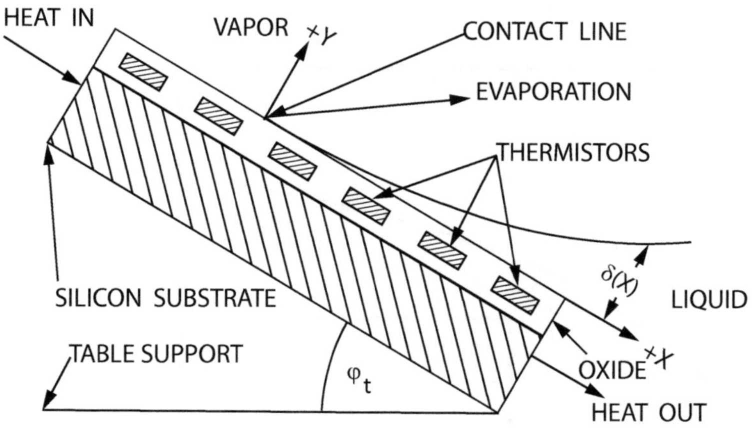

Fig. 4.23 [53] Schematic cross-section of thermistor array for use to generate a temperature gradient map at the liquid-gas interface during evaporation (©ECS 1989), reprinted with permission.

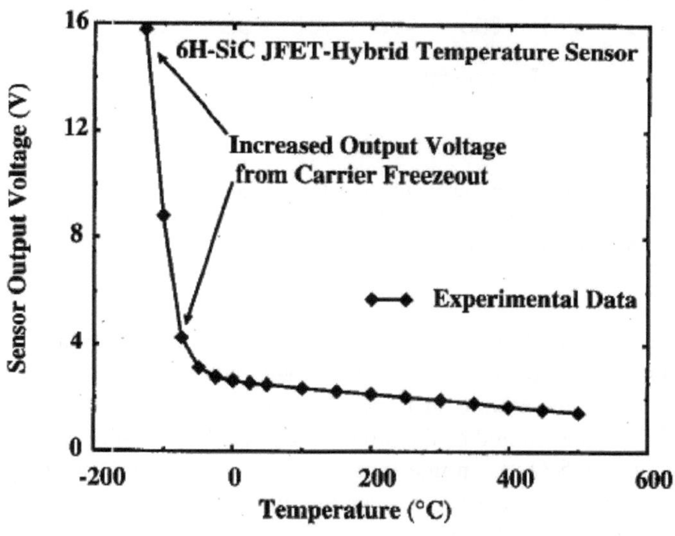

Fig. 4.24 [55] 6H-SiC junction gate field-effect transistor thermistor output voltage as a function of temperature (©Elsevier 1996), reprinted with permission.

near room temperature where the difference in CTE between Si and SiC is small, and the induced thermal strain (Figure 4.26) counteracts the strain due to stiffness softening [60]. In this regime of temperatures, frequency shift with temperature is very low. This provides a passive method capable of reducing thermal sensitivity for room temperature operation; however, at significantly higher temperatures, the sensitivity to temperature is amplified because of the CTE mismatch.

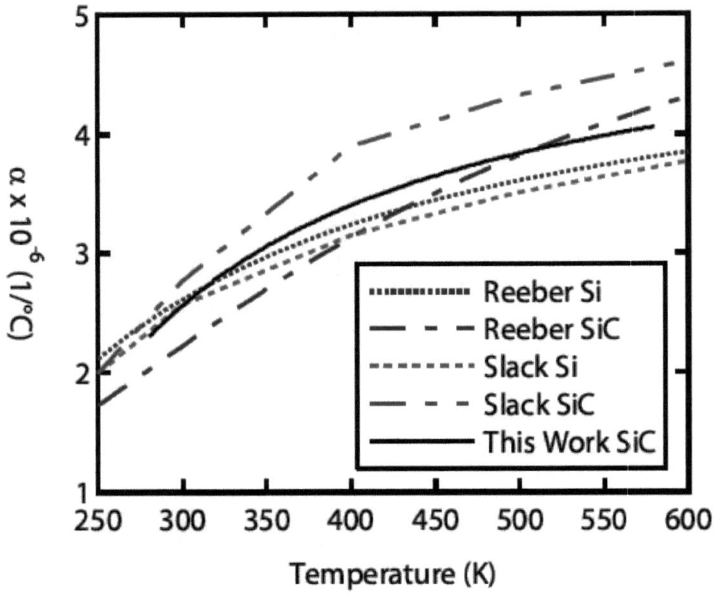

Fig. 4.25 [60] CTE as a function of temperature for 3C-SiC and Si as characterized by [57, 58, 59] and comparison to estimate of CTE for thin-film LPCVD 3C-SiC based on resonator response [60] (©IEEE 2009), reprinted with permission.

More extreme mismatches though tend to be detrimental to long-term survivability. Large cyclic loads are applied at the interfaces when materials with large CTE mismatch undergo large temperature changes. Therefore, whenever possible materials with closer CTE values should be the preferred design. For instance, it is better to use a-SiC compared to SiO_2 as an interlayer insulator between SiC or Si layers. For sustained high temperature operation, even the difference in CTE between the Si substrate, SiC device, and packaging layers can be problematic. Moving towards an all-SiC solution, utilizing a single-crystalline SiC substrate, alleviates the mismatch issue for high temperature operation. For very high temperature applications or applications that will continually cycle from room temperature to high operating temperatures, moving toward an all-SiC solution becomes a necessity for reliable long-term operation.

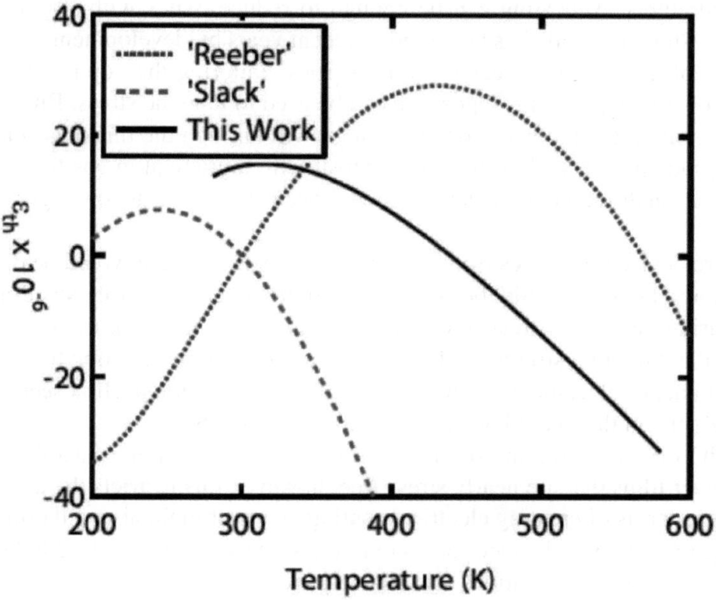

Fig. 4.26 [60] Thermal strain due to CTE mismatch between a silicon substrate and LPCVD 3C-SiC thin film based on literature for SiC and Si [57, 58, 59] and based on experimental measurements made using a double-anchored DETF resonator [15] ((©IEEE 2009), reprinted with permission.

4.3.2 Stress gradient and stress relaxation

For mechanical elements of microsystems, stress and stress gradients impact device performance. In extreme cases, films may curl out of plane upon release (Figure 2.35), which for many applications would render the device unusable. The films may delaminate from the surface. Thus, stress and stress gradient can be limiting factors in achievable film thickness. Especially for electrostatic-based structures, thicker films are desired to increase signal strength since for an in-plane operated structure, the forces generated along with the spring constant scale linearly with thickness as does the device mass. For example, if operating an in-plane resonator then increasing the thickness results in essentially no change in frequency or drive voltage; however, capacitive coupling to the sense electrodes also scales upwards.

As has been discussed previously, the mechanical properties of silicon carbide films are very stable with temperature. This is one of the reasons why it is a leading candidate material for harsh environment applications. But this also makes it difficult to anneal out a significant level of stress [61, 62]. With silicon dioxides and polysilicon layers, annealing often has a significant impact even when done at moderate temperatures. Annealing can be enough to reduce issues with stress or stress gradients. Silicon carbide has benefited in recent years by development of alternate methods of tailoring the mechanical properties. Tailoring the doping level, pressure, and flow of gases during deposition can be used to tune the stress. But for very thick films, often it is the variation in grain growth through the thickness that create a stress gradient. So work to likewise deposit with a gradient in gas flow or dopant concentration throughout the deposition has been able to create lower gradient films [63].

There still are device topologies from silicon micromachining that would be difficult to implement as is using SiC because they require a very long free-standing structure, requiring near-zero stress gradient. There is additional work to develop more strain gradient tolerant structures. By re-thinking some of the existing topologies in terms of anchor placement and how the structure folds upon itself is seen as an area of development that would be beneficial for SiC MEMS.

Epi-layer SiC or wafer-bonding of single-crystalline SiC to create SiCOI can create device layer films that are nearly stress free; however, this restricts the available topologies in terms of creating electrical routing layers or maintaining thermal budget if trying to integrate SiC electronics onto the same substrate as the MEMS device is important. So some solutions for the device realization standpoint need to be reviewed in the context of the overall system since they may limit options later in the system design.

4.3.3 Design for high g-shock survivability

Some level of shock survivability is usually considered as part of a commercial MEMS sensor mechanical design, from intentional accelerations expected in the field to accidental events such as dropping the sensor from a table. Although MEMS, due to their low inertia, are innately robust to moderate g-shock, certain harsh environment applications such as munitions monitoring require surviving shock events of 100,000 g or more. SiC has a favorable stiffness to weight ratio and good yield strength, but at these very high shock levels, more traditional designs may need to be modified to ensure device survivability. To highlight how design can be used to improve shock survivability, an example of how a SiC comb-driven double-ended tuning fork (CDDETF) resonant sensor is modified for high g-shock survivability will be discussed [15].

CDDETF resonators are commonly used in MEMS sensor applications and utilize electrostatic comb arrays for drive and sense of tine deflection (Figure 4.27a). The comb-array provides a higher transduction constant (due to increased capacitor surface area), which does not vary with the magnitude of tine excitation (minimizing distortion of the output signal). In addition, air damping is less for a comb array than for a gap-closing capacitor array of equivalent transduction efficiency. Thus, even though the added mass of the comb array decreases the resonator sensitivity

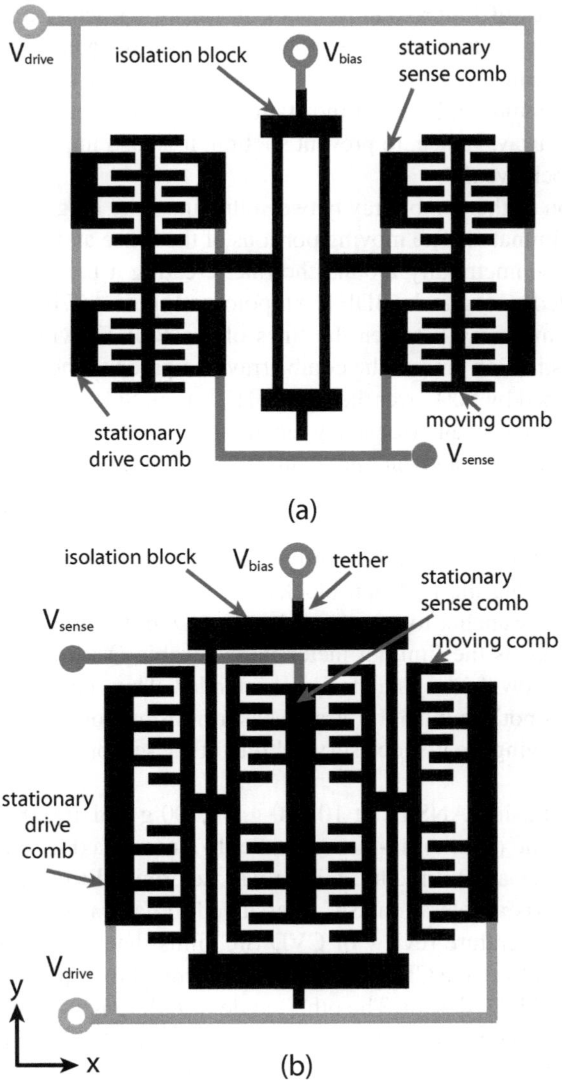

Fig. 4.27 [15] Schematics of (a) standard comb-driven double-ended tuning fork (CDDETF) resonator and (b) balanced-mass double-ended tuning fork (BDETF). (©IEEE 2007), reprinted with permission.

to applied strain or force [64], this is often tolerated because of the superior electrical characteristics achieved with the electrostatic comb array. This large structure is suspended far from the resonating tine and is detrimental to high shock survivability. It leads to a structure that is susceptible to deflection under shock, and the resulting load it creates on the tine and anchors cause high local stresses on other portions of the structure.

There are two primary modes of failure due to shock in MEMS: fracture and stiction. Preventing fracture, obviously, requires management of strains generated during a shock event. Stiction can occur either between the device layer and the substrate or, as is much more common, between the stationary and moving comb fingers of the capacitive comb array. Hence, to prevent stiction, it is also important to limit deflection during a shock event.

By folding the sense portion of the comb array between the tines, the long lever of the CDDETF structure is eliminated. The moving portions of the sense and drive comb arrays are now placed symmetrically around the tine, creating a balanced-mass comb-driven double-ended tuning fork (BDETF) topology (Figure 4.27b). To facilitate folding the sense comb array between the tines of the BDETF without significantly reducing the transduction area of the comb array compared to the CD-DETF, the tine length is increased by 50% over the CDDETF design. Since the primary flexural mode resonant frequency and frequency sensitivity to applied strain of a clamped-clamped tine is constant for a given tine width-to-length ratio, the width of the BDETF tine is proportionally increased so as to maintain the same resonant frequency and sensitivity. In order to fit one half of the comb array between the tines requires widening the isolation block significantly over the CDDETF configuration. It is possible that widening the isolation blocks that connect the tines to the substrate anchors may increase anchor loss compared to the CDDETF; however, fluidic damping, not anchor loss, is the limiting factor in achievable Q for a resonator operated in air, which is how the original design is intended. Thus, changing the resonator to a BDETF does not diminish the strain sensor transduction performance while theoretically improving the g-shock survivability of the sensor over the standard CDDETF.

Simulations were conducted using ANSYS at 10,000 g, 50,000 g and 100,000 g in three directions: out-of-plane (z-axis), in-plane parallel to tines (y-axis), and in-plane perpendicular to tines (x-axis). Fracture failure was determined to occur first for both designs under z-axis acceleration. SiC is assumed to have a fracture strain limit of 0.17% based on literature review of CVD SiC films that examined comparably sized features. Therefore, the CDDETF is fracture-limited to 70,000 g, compared to 94,000 g for the BDETF design. The other mode of failure is stiction. For this type of structure, contact occurs either with the substrate or with the moving comb array fingers contacting the stationary fingers. Contact with the substrate will occur under 23,000 g z-axis acceleration for the CDDETF; whereas, the BDETF can survive 150,000 g based on extrapolation of the simulation data. Comb finger contact occurs at 93,000 g y-axis acceleration for the CDDETF while the BDETF reduces deflection in the y-axis by 98%.

By understanding the failure modes of the device structure along with general design constraints of the sensor enables sensor redesign that increases shock survivability considerably (in this example as much as four-fold). With careful design considerations, these gains can be made without significant compromise to sensor performance. Sometimes, as in this particular example, understanding the dominant loss mechanisms in the device allows improving shock survivability without reducing sensor sensitivity.

References

1. Gao D, Wijesundara MBJ, Carraro C, Howe RT, Maboudian R (2004). Transformer coupled plasma etching of 3C-SiC films using fluorinatedchemistry for microelectromechanical systems applications. Journal of Vacuum Science Technology B 22(2):513–518
2. Jones DG, Pisano AP (2010). Aluminum nitride as a masking material for the plasma etching of silicon carbide structures. IEEE 23rd International Conference on Micro Electro Mechanical Systems, Hong Kong, Jan. 24-28:352–355
3. Myers DR, Cheng KB, Jamshidi B, Azevedo RG, Senesky DG, Wijesundara MBJ Pisano AP (2009). A Silicon Carbide Resonant Tuning Fork for Micro-Sensing Applications in High Temperature and High G-Shock Environment. Journal of Micro/Nanolithography, MEMS, and MOEMS 8:021116
4. Suster M, Guo J, Chaimanonart N, Ko WH, Young DJ (2006). A High-Performance MEMS Capacitive Strain Sensing Microsystem. JMEMS 15(5):1069–1077
5. Hetherington DL, Sniegowski JJ (1998). Improved Polysilicon Surface-micromachined Micromirror Devices using Chemical-mechanical Polishing. SPIE's 43rd Annual Meeting, San Diego, CA, July 22, 1998.
6. Fu XA, Dunning J, Zorman CA, Mehregany M (2005). Polycrystalline 3C-SiC thin films deposited by dual precursor LPCVD for MEMS applications. Sensors and Actuators A 119:169–176
7. Rogacheva NN (2004). The Theory of Piezoelectric Shells and Plates. CRC Press.
8. Toriyama T (2004). Piezoresistance Consideration on n-type 6H SiC for MEMS-based Piezoresistance Sensors. J. Micromech. Microeng. 14:1445–1448
9. Azevedo RG, Chen I-Y, OŔeilly OM, Pisano AP (2005). Influence of Sensor Substrate Geometry on the Sensitivity of MEMS Micro-extensometers. IMECE Orlando, FL:IMECE2005-82724
10. MicroStrain, Inc. USA. http://microstrain.com
11. Beckwith TG, Marangoni RD, Lienhard JH (1993). Mechanical Measurements. Addison-Wesley, 5th Ed.
12. Jamshidi B, Azevedo RG, Wijesundara MBJ, Pisano AP (2007). Corrosion Enhanced Capacitive Strain Gauge at 370C. MEMS 2007, 20th IEEE SENSORS 2007 Conference on Micro and Nano sensors Technical Digest 2007:804–807
13. Tongue BH (2001). Principles of Vibration. Oxford University Press, 2nd Ed.
14. Azevedo RG, Zhang J, Jones DG, Myers DR, Jog AV, Jamshidi B, Wijesundara MBJ, Maboudian R, Pisano AP (2007). Silicon Carbide Coated MEMS Strain Sensor for Harsh Environment Applications, MEMS 2007, 20th IEEE International Conference on Micro Electro Mechanical Systems, Technical Digest 2007:643–646
15. Azevedo RG, Jones DG, Jog AV, Jamshidi B, Myers DR, Chen L, Fu, XA, Mehregany, M, Wijesundara, MBJ, Pisano, AP (2007). A SiC MEMS Resonant Strain Sensor for Harsh Environment Applications, IEEE Sensors, 7(4):568–576
16. Wojciechowski KE, Boser BE, Pisano AP (2004). A MEMS resonant strain sensor operated in air. MEMS Conference 2004: 841–845
17. Krauthammer T, Venstal E (2001). Thin Plates and Shells: Theory, Analysis, and Applications. CRC Press.
18. Di Giovanni M (1982). Flat and Corrugated Diaphragm Design Handbook. CRC Press.
19. Beeby S (2004). MEMS Mechanical Sensors. Artech.
20. Zeirmann R, von Berg J, Reichert W, Obermeiȩr E, Eickhoff M, Krotz G (1997). A high temperature pressure sensor with β-SiC on SOI substrates. Int. Conf. Solid State Sensors and Actuators, Chicago, June 16-19:1411–1414
21. Okojie R, Ned A, Kurtz A (1997). Operation of 6H-SiC Pressure Sensor at 500 °C. Tech. Dig. 1997 Int. Conf. Solid State Sensors and Actuators,Chicago IL, June 1619:1407-1409
22. Eickhoff M, Möller H, Krotez G, Berg JV, Ziermann R (1999). A High Temperature Pressure Sensor Prepared by Selective Deposition of Cubic Silicon Carbide on SOI Substrates. Sensors and Actuators A 74:56–59

23. Faris W, Mohammed H (2004). A comparison between two solution approaches for diaphragm-based capacitive pressure microsensor. ICSE Dec 7-9:295–297
24. Pakula LS, Yang H, Pham HTM, French PJ, Sarro PM (2004). Fabrication of a CMOS compatible pressure sensor for harsh environments. J. Micromech. and Microeng. 14(11):1478–1483
25. Young DJ, Du J, Zorman CA, Ko WH (2004). High-Temperature Single-Crystal 3C-SiC Capacitive Pressure Sensor. IEEE Sensors Journal 4(4):464–470
26. Chen L, Mehregany M (2008). A silicon carbide capacitive pressure sensor for in-cylinder pressure measurement. Sensors and Actuators A 145146:2-8
27. Seshia AA, Palaniapan M, Roessig TA, Howe RT, Gooch RW, Schimert TR, Montague S (2002). A vacuum packaged surface micromachined resonant accelerometer. JMEMS 11(6):784–793
28. Atwell AR, Okokie RS, Kornegay KT, Roberson SL, Beliveau A (2003). Simulation, fabrication and testing of bulk micromachined 6H-SiC high-g piezoresistive accelerometers. Sensors and Acutators A 104:11–18
29. Rajgopal S, Zula D, Garverick S, Mehregany M (2009). A Silicon Carbide Accelerometer for Extreme Environment Applications. Materials Science Forum 600-603:859–862
30. Pakula LS, Yang H, French PJ (2003). A CMOS compatible SiC accelerometer. Sensors 2003:761–764
31. Fleischman AJ, Roy S, Zorman CA, Mehregany M (1996). Polycrystalline silicon carbide for surface micromachining. MEMS 1996, San Diego, Feb. 11-15:234-238
32. Fleischman AJ, Roy S, Zorman CA, Mehregany M (1997). Behavior of polycrystalline SiC and Si surface-micromachined lateral resonant structures at elevated temperatures. Int. Conf. Silicon Carbide, IIINitrides, and Related Materials, Stockholm, Aug. 31-Sept. 5:643–644
33. Wiser R, Zorman ZA, Mehregany M (2003). Fabrication and testing of vertically-actuated polycrystalline silicon carbide micromechanical resonators for MHz frequency applications. Transducers 2003, Boston, June 8-12:1164–1167
34. Bhave SA, Gao D, Maboudian R, Howe RT (2005). Fully-differentical poly-SiC lame-mode resonator and checkerboard filter. MEMS 2005, 18th IEEE International Conference on Micro Electro Mechanical Systems, Technical Digest (2005):223–226
35. Jiang L, Cheung R, Hedley J, Hassan M, Harris AJ, Burdess JS, Mehregany M, Zorman CA (2006). SiC cantilever resonators with electrothermal actuation. Sensors and Actuators A 128:376–386
36. Seshia AA, Howe RT, Montague S (2002). A Micromechanical Resonant Output Gyroscope. Proc. IEEE MEMS 2002, Las Vegas, Jan. 20-24:722–727
37. Xie H, Fedder GK (2003). Fabrication, Characterization, and Analysis of a DRIE CMOS-MEMS Gyroscope. IEEE Sensors Journal 3(5):622–631
38. Maenaka K, Ioku S, Fujita T, Takayama Y (2005). Design, fabrication and operation of MEMS gimbal gyroscope. Sensors and Actuators A 121(1):6–15
39. Cheung R (ed.) (2006). Silicon Carbide Micro Electromechanical Systems. Imperial College Press.
40. Lloyd Spetz A, Baranzahi A, Tobias P, Lundström I (1997). High temperature sensors based on metal-insulator-silicon carbide devices. Phys. Stat. Sol. A 162:493–511
41. Baranzahi A, Lloyd Spetz A, Glavmo M, Carlsson C, Nytomt J, Salomonsson P, Jobson E, Häggendal B, Martensson P, Lundström I (1997). Response of metal-oxide-silicon carbide sensors to simulated and real exhaust gases. Sensors and Actuators B 43:52–59
42. Moritz W, Fillipov V, Vasiliev A, Terentjev A (1999). Silicon carbide based semiconductor sensor for the detection of fluorocarbons. Sensors and Actuators B 58:486–490
43. Svennigstorp H, Widén B, Salomonsson P, Ekedahl L-G, Lundström I, Tobias P, Lloyd Spetz A (2001). Detection of HC in exhaust gases by an array of MISiC sensors. Sensors and Actuators B 77:177–185
44. Wright NG, Horsfall AB (2007). SiC sensors: a review. J. Phys. D: Appl. Phys. 40:6345–6354
45. Wiche G, Berns A, Steffes H, Obermeier E (2005). Thermal analysis of silicon carbide based micro hotplates for metal oxide gas sensors. Sensors and Actuators A 123-124:12–17

46. Cree Inc., USA. http://www.cree.com
47. Zhu H, Chen X, Cai J, Wu Z (2009). 4HSiC ultraviolet avalanche photodetectors with low breakdown voltage and high gain. Solid-State Electronics 53(1):7–10
48. Torvik JT, Pankove JI, Van Zeghbroeck BJ (1999). IEEE Trans. Electron Devices 46(7):1326–1331
49. Seely JF, Kjorntattanawanich B, Holland GE, Korde R (2005). Response of a SiC photodiode to extreme ultraviolet through visible radiation. Optics Letters 30(23):3120–3122
50. Strokan NB, Ivanov AM, Savkina NS, Streichuk AM, Lebedev AA, Syväjärvi M, Yakimova R (2003). Detection of strongly and weakly ionizing radiation by triode structure based on SiC films. J. Appl. Phys. 93:5714–5719
51. Nagai T, Yamamoto K, Kobayashi I (1982). SiC thin-film thermistor. J. Phys. E: Sci. Instrum. 15:520–524
52. de Vasconcelos EA, Khan SA, Zhang WY, Uchida H, Katsube T (2000). Highly sensitive thermistors based on high-purity polycrystalline cubic silicon carbide. Sensors and Actuators 83:167–171
53. Kiewra EW, Wayner Jr PC (1989). The Development of a Thin-Film Silicon Carbide Thermistor Array for Determining Temperature Profiles in an Evaporating Liquid Film. J. Electrochem. Soc. 136(3):740–744
54. Casady JB, Dillard WC, Johnson RW, Rao U (1996). A Hybrid 6H-SiC Temperature Sensor Operational from 25 °C to 500 °C. IEEE Trans. Comp., Pack., Manuf. Tech. A 19(3):416–422
55. Casady JB, Johnson RW (1996). Status of Silicon Carbide (SiC) as a Wide-Bandgap Semiconductor for High-Temperature Applications: A Review. Solid-State Electronics 39(10):1409–1422
56. Riza NA, Sheikh M (2010). Silicon Carbide-Based Extreme Environment Hybrid Design Temperature Sensor Using Optical Pyrometry and Laser Interferometry. IEEE Sensors Journal 10(2):219–224
57. Slack GA (1975). Thermal expansion of some diamondlike crystals. J. Applied Physics 46(1):89–98
58. Reeber RR, Wang K (1996) Thermal expansion and lattice parameters of group IV semiconductors. Materials Chemistry and Physics 46:259-264
59. Reeber RR, Wang K (1996) Thermal expansion of β-SiC, GaP and InP. Materials Research Symposium Proceedings 410:211-216
60. Azevedo RG, Myers DR, Pisano AP (2009). Temperature-insensitive silicon carbide resonant micro-extensometers. Transducers 2009, Denver, June 21-25:268–271
61. Berry BS, Pritchett WC, Fuentes RI, Babich I (1991). Stress and thermal expansion of β-SiC films by the vibrating-membrane method. J. Mater. Res. 6(5):1061–1065
62. Okojie RS (2004). Inelastic Stress Relaxation in Single Crystal SiC Substrates. Material Science Forum 457-460:375–378
63. Zhang J, Howe RT, Maboudian R (2006). Control of strain gradient in doped polycrystalline silicon carbide films through tailored doping. J. Micromech. Microeng. 16:L1-L5
64. Roessig TA, Howe RT, Pisano AP, Smith JH (2007). Surface-micromachined resonant accelerometer. 1997 International Conference on Solid-State Sensors and Actuators:859–862
65. Srikar VT, Senturia SD (2001). The design and analysis of shock resistant microsystems (MEMS). Transducers 2001, Munich, Germany:1370–1373

Chapter 5
Packaging

Packaging is typically required to provide some level of hermeticity to the sensor and electronics. Without this protection, the sensor or electronics performance would degrade or drift, aliasing the output characteristics and potentially leading to premature failure of the device. These issues are compounded for harsh environment applications. Highly corrosive media require highly corrosion resistant materials be used for packaging, limiting the available material set that can be used. High temperature environments increase the rate of corrosion and diffusion as well as can decrease fatigue life or may simply exceed the melting point or glass transition point of certain common packaging materials. It also can introduce significant internal stresses due to mismatch in thermal expansion rates of the various materials inside the package. High pressure and high shock environments additionally require components be properly sized or a different mechanical topology implemented so that they can survive the high mechanical forces encountered.

Sensor packaging rarely gets to be as simple as casing the entire sensor in a rugged box. Interconnects through the package to the environment are nearly always needed. Although this typically is only an electric route for either signal or power in and out of the package, for non-inertial sensors this also includes a force transduction port such as a pressure tube or gas port. Accounting for these interconnects as well as linking multiple packaged components together to form the entire sensing system create weak points in the package design or leads to very costly solutions.

Hermeticity applies to preventing ingress or escape of air. Although as will be discussed this is sometimes not the critical issue. In certain cases, prevention of liquid ingress is sufficient. Some devices are ruined by accumulation of moisture in the package which can cause electrical shorting, electrochemical corrosion, or stiction in narrow-gap mechanical structures. In the harsh environment context, environmental liquids and gases may themselves be significant etchants of the encapsulated structures, which adds another failure mechanism.

Materials that prevent air and water moisture ingression range from single-crystal to metals to polycrystalline films. Although the materials themselves may not directly allow gas to pass through because of their high density and low diffusivity,

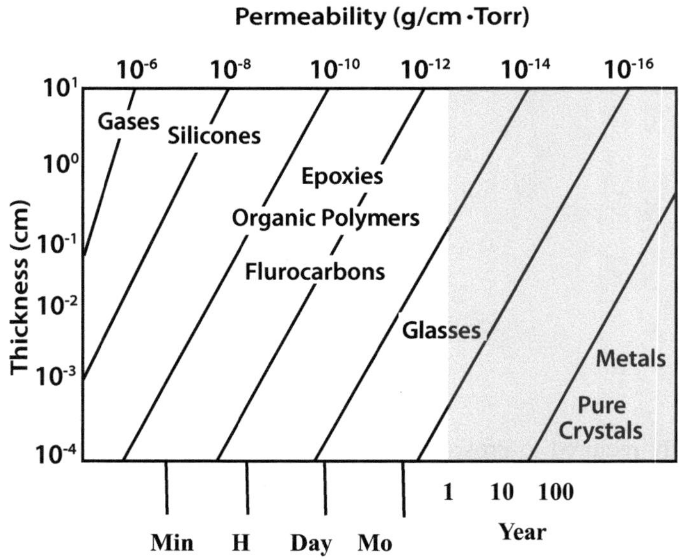

Fig. 5.1 [1] Calculated permeability of water through organic and inorganic materials (©Springer 1997), reprinted with permission. Since device lifetimes typically need to be on the order of years (grayed region), this generally excludes most polymer alternatives.

gases can make their way into the package by traveling along grain boundaries. This is especially problematic in high-temperature and high-pressure environments since both of these aid in diffusion through the packaging material. Polymers easily uptake water in comparison, making them often not suitable for long-term hermetic applications. Figure 5.1 shows diffusion rates through various materials as a function of thickness.

Since MEMS research into harsh environment packaging is to date relatively limited, this chapter will provide an overview of the current state of packaging technologies for MEMS in general. This is to provide the reader with an appropriate context for the limited subset of packaging techniques that will be applicable to harsh environment applications. The various environments that make up the harsh environment space will be examined in the context of MEMS packaging techniques available and packaging techniques that are not suitable for that subset environment will be detailed. Techniques that fit the necessary criteria will be reviewed or needs anticipated when current research has not yet explored that regime.

5.1 Zero-level MEMS Packaging

MEMS packaging is in many ways an extension of IC packaging and borrows many of the same techniques. MEMS packaging though is complicated for many applica-

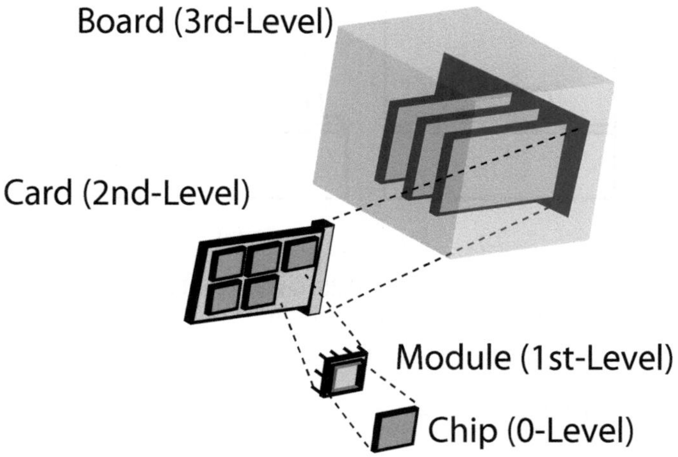

Fig. 5.2 Schematic illustrating the nomenclature for the various levels of packaging and backend integration that goes from interfacing a die to a system.

tions by two additional requirements: 1) the need to provide an internal cavity for a free-standing mechanical device that is used as a sense element and 2) the need to provide not only electrical but physical (*e.g.*, fluidic, thermal) interaction with the environment outside the package. Because the sense element is formed at the die level, this places a significant effort on developing robust 1st- and 0-level packaging (Figure 5.2) techniques to protect these fragile mechanical structures while providing the necessary environmental coupling. The need for environmental coupling is particularly troublesome in a harsh environment context since it often requires 0-level packaging solutions that are highly corrosion resistant when a pure IC may be able to push such requirement off until the 1st-level package.

Traditional MEMS research has developed a wide range of 0-level packaging schemes because they tend to be tailored to the sensor type and sensor fabrication technique. These techniques generally differ by material set but can be grouped logically into form-factor and attachment process. Techniques for 0-level packaging can first be divided into two broad classes: 1) bonding techniques, in which some lid with cavity is placed over the MEMS device and bonded or 2) deposition techniques, in which additional layers are added over the micro-structure layer to form a lid (Figure 5.3a). Bonding techniques can be further divided into wafer-to-wafer (Figure 5.3b) and transferred lid (Figure 5.3c) techniques.

Harsh environment packaging further utilizes corrosion resistant coatings. This allows using traditional materials such as silicon that are mechanically robust and utilize well-characterized processing techniques to create the packaging structure. By depositing thin-film SiC over these conventional structures allows them to be used in chemically corrosive or oxidative environments at minimal added cost.

These techniques, along with their advantages and disadvantages, will be laid out. Integration and routing of electrical interconnects as they pertain to 0-level packaging will also be discussed.

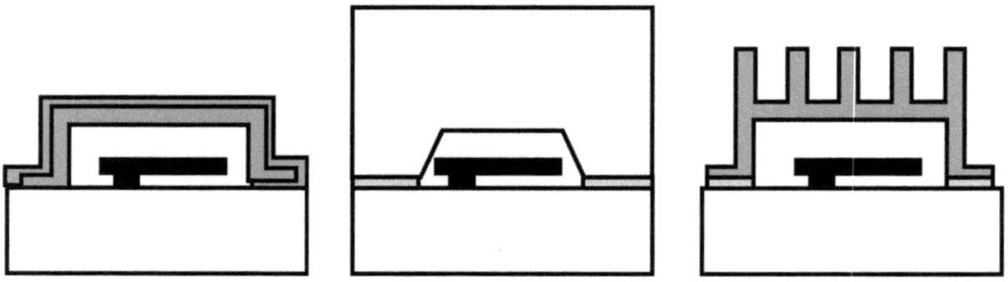

Fig. 5.3 Schematic examples of 0-level packaging for MEMS. (a) Deposited cap, fabricated in a monolithic process, (b) bulk wafer cap, (c) transferred hexsil rib cap.

5.1.1 Bonding Techniques

One major class of packaging techniques involves either sealing a cavity in the device wafer or creating a cavity in a second wafer or die and bonding the two wafers or wafer and die together. Bonding includes techniques that introduce a separate bond material to facilitate the bonding process either by modifying the process temperature to form the bond or by providing a medium that provides the necessary adhesion. Bonding can in some cases be achieved without a separate adhesion layer as well.

5.1.1.1 Wafer-to-wafer

Wafer-to-wafer bonding is a common technique in silicon MEMS processing. It has been utilized to create a variety of sacrificial and device layer stacks for device fabrication (*e.g.* SOI wafer technology). It is also a useful packaging technique as well. It gained wide use as a packaging technique for pressure sensors. It is widely utilized to provide an encapsulated vacuum or controlled atmosphere cavity as a pressure reference in pressure sensor fabrication. A variety of bonding techniques have been developed. They will be briefly reviewed here, although as will be pointed out they are not all suitable for high temperature or harsh environment applications.

The most common method of encapsulation is still bulk wafer capping. This method is implemented by etching recesses into a second wafer, which is bonded to the device layer through an intermediate bond material or by anodic or thermal compression bonding. This method is robust and reliable; however, it requires a relatively large bond ring, which consumes considerable real estate on the die. It also requires a highly planar surface at the bond ring area on both the cap and device wafer. This is not always compatible with electrical routing or interconnects, although routing processes can be tailored to this type of packaging in some cases.

Anodic bonding works well for microfluidics because it requires the use of a glass and silicon wafer pair. This allows for optical interrogation of fluidic channels if necessary. Anodic bonding occurs through migration of positive ions from the glass substrate to the silicon-glass interface under an applied electric field (200-1500 V) and is assisted through the use of vacuum and elevated temperature (approximately 400 °C). CTE mismatch between the wafers can cause considerable residual stress and wafer bow; however, glass manufacturers now offer specialty glass substrates that are fairly well matched to silicon up to several hundreds of degrees Celsius [2]. The anodic bonding method tends to create trapped charge regions at the wafer interface, which can cause degradation of electronics performance. A variation of this method is accomplished by growing an oxide layer on a silicon wafer to fuse two silicon wafers together. This is one method of bonding that can be used to form SOI wafers. This method has been adapted to make SiC-on-insulator substrates [3, 4], although to date it has not been used for harsh environment sensor packaging.

Thermal fusion bonding is another method commonly used to bond two silicon wafers together. This bonding method uses very high temperatures to create interdiffusion between the wafers. This method does not use a bonding layer and requires very clean and smooth surfaces on both silicon substrate and silicon capping wafer as well as processing temperatures upwards of 1000 °C to achieve sufficient interdiffusion to bond the wafers together. Therefore, this type of bonding exceeds the thermal budget for silicon IC devices. Although this is a very strong bond method, it requires very smooth surfaces and high cleanliness is achieved using chemicals that can etch typical IC dielectric and metal layers. So this method is usually limited to initial substrate preparation, prior to IC fabrication.

This type of high temperature diffusion bonding can also be used to form SOI wafers, which allows bonding without special equipment to apply a high voltage. This is primarily a matter of convenience and throughput since high capacity tube furnaces can be used and may already exist at a given fabrication facility for high temperature annealing. A variant of this type of processing has been successfully utilized to form high quality SiC-on-insulator substrates for microwave electronics [5]. Although, like with anodic bonding, the technique may work for packaging harsh environment sensors, to date no reports are found in the literature.

To depress the bonding temperature, a wide variety of adhesion layers have been introduced to wafer bonding. The adhesion layer can have a relatively low melting temperature, which can create low temperature bonds (*e.g.*, Sn, In, BCB, Parylene). With thicker bond layers, this adhesive layer approach also reduces the need for a highly planar surface [6]. It is often advantageous to use a material that will form a eutectic alloy with silicon (*e.g.* Au [7] or Al [8]). Eutectic alloys of the proper ratio can be formed by melting one or both materials (usually metals). The eutectic point allows for melting of the alloy at a temperature well below the melting temperature of either of the constituents. Another alternative is to use a multi-material bond ring stack of pre-deposited materials that create a eutectic bond pair (*e.g.* Au-Sn [9], Pb-Sn [10] or In-Sn [11]).

Although these bond materials can greatly reduce bonding temperature, this likewise limits operating temperatures to prevent yield or creep of the bond. Additionally, some of these bond materials are polymers or metals that are easily etched, making them less suitable for corrosive or oxidizing environments; however, if limited lifetime is needed, some compromise between bond ring width and etch rate may be feasible. Additionally, if high temperature operation is required, metals and even silicon dioxide are not well matched to SiC in terms of CTE. Characterizing fatigue life of these types of composite structures under thermal cycling up to 300 or 600 °C is an open area for research.

Wafer-to-wafer packaging techniques are becoming increasingly common in silicon sensor commercial packaging. The current drawback for researching these techniques for SiC is the relative cost of the wafer substrate. A 100mm diameter substrate of pyrex or silicon can be had for $10 to $20 per wafer. SiC substrates are approximately 200 times the cost of silicon currently. This has pushed 0-level packaging research towards structural coatings of SiC or chemical passivation of silicon for packaging. It is expected that as the price of SiC substrates decrease, wafer-to-wafer packaging has a strong potential as a MEMS-first configuration that preserves a high-quality SiC layer for IC fabrication.

5.1.1.2 Transferred Lid Technique

Transferred lid capping is a method that allows formation of a lid in its own fabrication process completely separate from the device wafer before it is finally transferred to the device wafer. This allows for a compromise that decouples the fabrication process of the lid structure but can preserve a high precision alignment between all the lids and devices on a wafer. Thus, it is possible to transfer all the lids to the devices at once instead of in a serial process.

Although it is possible that transferred lid capping could be used to create a thick film lid, it has primarily been demonstrated for creating a lid using a thin-film molding process in the lid substrate. In a demonstration for SiGe 0-level encapsulation, hexsil ribs were added to the cap through etching of narrow trenches along the lid span in order to stiffen the resulting membrane encapsulation [12, 13]. The ribs make even a thin film mechanically strong enough to support an atmosphere difference in pressure across the membrane. Transferred lid methods also require a bond ring around the device, but tend to utilize a much smaller bond ring area than the bulk wafer capping method.

Because these techniques either utilize a bond ring material or are formed from the substrate typically using deep etch techniques, this in turn limits how far one can shrink the bond ring. In addition, for methods that use semi-hermetic materials such as polymers, the time the package can protect the encapsulated devices from moisture is a strong function of the width of the bond ring. Hence, narrow polymer bond rings are detrimental to long-term use. As discussed previously, polymers would not be suitable for many chemical or high temperature environments.

Existing literature has been focused on packaging silicon circuitry. Hence, very low temperature melting metals have been used. But if the same concept were applied to SiC electronics, higher melting point bond materials such as gold may be feasible and may be compatible with moderate to high temperature operation.

5.1.2 Deposition Techniques

An alternative to the bonded encapsulation methods discussed in the previous section is to deposit the encapsulation directly over the MEMS structure. Deposition techniques involve depositing at least one, but often multiple, layer after the MEMS structure is defined. Often these techniques are based on a second device layer technique in the sense that a sacrificial layer is deposited before a packaging structural layer is deposited. This is typically done before the micro-sensor is released so as to avoid recoating or re-attaching the mechanical structure to the substrate.

Another popular technique for chemically harsh environments is to passivate a non-corrosion resistant set of materials by depositing a very thin, conformal coating of LPCVD or PECVD SiC over the surfaces that will be exposed to the corrosive environment. Although this does not alleviate temperature constraints of the underlying structure, corrosion protection is enhanced, allowing an otherwise standard material set to be used for applications where corrosion limits the useful life of these devices. Both the structural scaffold and conformal coating techniques will be discussed in more detail next.

5.1.2.1 Structural Thin-films

Structural thin-films as 0-level packaging aim to reduce the overall footprint of the package and shrink the final device dimensions while utilizing a parallel processing technique. If yields of the packaging technique are very high, this has the potential to decrease overall device cost. Some techniques further aim to either utilize processes for the packaging with very low temperatures to be compatible with MEMS-last IC integration or finish processing with a high quality, highly planar silicon surface to be compatible with MEMS-first IC integration.

The generic structural thin-film approach requires first encasing the MEMS structure with a sacrificial material such as silicon dioxide or photoresist. This provides a stand-off between the moving device and encapsulation. Then the thin-film structural layer of the encapsulation (also known as the scaffold layer) is deposited over the sacrificial material and photolithographically patterned with either side [14, 15] or top etch access holes [16, 17]. An etchant, typically liquid but possibly chemical vapor, is then used to remove the sacrificial material. If the packaging sacrificial material is different from the micro-structure sacrificial material, a second etchant may be used to fully release the free-standing features of the device layer. As mentioned in the device fabrication flow in Chapter 4, a critical point drying step may be introduced after the etch to prevent stiction of the device and scaffold layers due to liquid meniscus forces during drying. A sealing film is then put down to seal the etch access holes. This film could be a spun-on polymer [18], but is more typically a deposited metal or ceramic film. With few exceptions, this sealing film will need to be patterned to define electrical contact pads in a conductive sealing film or clear away a non-conductive sealing film to create a conductive path to underlying electrical contacts.

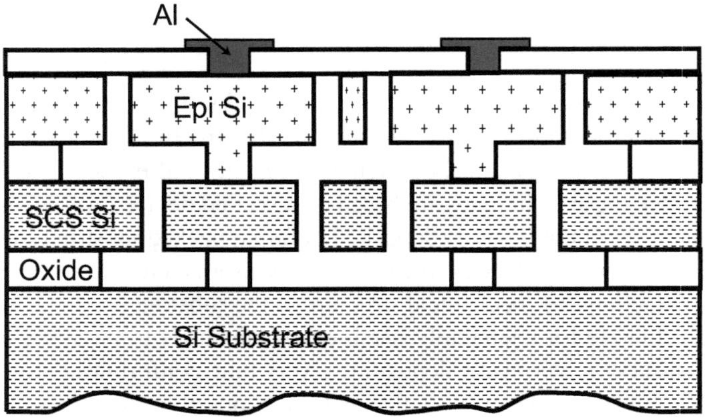

Fig. 5.4 [16] Schematic of the process cross-section of an encapsulated beam resonator using the epitaxial silicon scaffold layer technique (©IEEE 2003), reprinted with permission.

Because of both permeability concerns and restrictions on the practical thickness range of deposited films, deposited packages for devices requiring long lifetimes are commonly made from metals [15]. Poly-crystalline films are also used [14, 17]. Single-crystalline materials would also be a good choice because reducing grain boundaries reduces the diffusion paths through the film, although in practice it is difficult to create a single-crystalline deposited film.

A high-yield example of this thin-film technique has been developed jointly between Stanford University and Robert Bosch, GmbH and later adopted and further refined by SiTime, Inc., a start-up developing silicon mechanical resonators as replacements for crystal quartz timing references [16]. The process aims to fabricate the mechanical micro-structures first but end with a high quality silicon layer for post-MEMS IC integration. A previously published version of this process starts with a SOI wafer with a device layer tens of microns thick (Figure 5.4). The silicon is etched to define the resonator structure. This is followed by several microns of silicon dioxide so as to cover and span the trenches. The oxide is planarized and patterned to create electrical vias. A thick epitaxial silicon layer is grown as the scaffold layer, typically 20 to 40 microns thick at temperatures of approximately 1200 °C. This thickness is set by a deflection limit of the largest span distance that can withstand the pressure of injection molding during 1st-level packaging. The scaffold layer is again patterned to define the electrical vias and etch access holes. A vapor hydrofluoric acid etch is used to remove the device and scaffold layer sacrificial oxides. Finally, another silicon dioxide layer is deposited to seal the etch access holes and capture the deposition vacuum inside the resonator cavity, which is beneficial in reducing resonator damping (increasing Q-factor). In the device-only version of this process the sealing oxide is etched at the electrical via features and a patterned metal is added to reduce contact resistance and promote wire-bonding (Figure 5.5).

This process utilizes both rectangular etch access holes and purposely offsetting the placement of these holes from being directly over the resonator structure when-

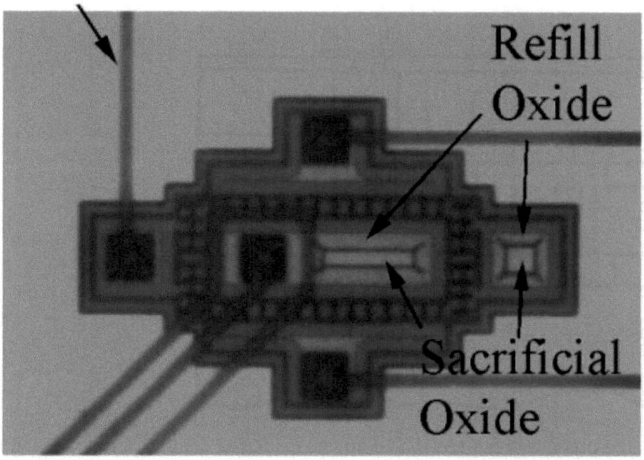

Fig. 5.5 [16] Top view IR micrograph of finished resonator structure (©IEEE 2003), reprinted with permission.

ever possible to minimize mass loading of the resonator by the subsequent sealing layer deposition. Offsetting the holes prevents initial deposition from having a line-of-sight avenue to the resonator. The rectangular holes allows for effective etching of a hole with a minimum dimension smaller than a square hole can be created in the same process because the etch process is area dependent. This allows reducing the necessary sealing oxide thickness to pinch off the etch holes because this is minimum dimension dependent.

This process was further refined to allow for direct CMOS integration on the device substrate (Figure 5.6). The initial scaffold layer is reduced to only 1-2 μm thick and instead sealed with a thick epitaxial silicon growth (SiTime's EpiSeal^{TM} process) after removal of the sacrificial oxide [19]. By sealing the cavities with a thick silicon instead of a thin silicon dioxide, the packaged device can be planarized again after the sealing process to reveal a flat, high quality silicon surface for CMOS fabrication without concern about re-opening the etch access holes (Figure 5.7).

This type of encapsulation process for a microsensor is adaptable to SiC. SiCOI wafers have been demonstrated with a silicon handle, but the process should be readily transferable to using a SiC handle wafer. LPCVD films have been demonstrated at 7 to 8 μm [20]. The higher elastic modulus of SiC compared to silicon means a reduced thickness can be used, although increased film thickness is preferable. The larger challenge however is that high aspect ratio etching is needed to create small etch access holes through this thick film [21]. Part of the challenge is battling mask erosion, which leads to a widening of the etch hole. Current etching techniques utilizing metal masking materials such as Ni [22, 23] or alternative dielectrics to silicon dioxide such as AlN [24] may enable development of a high aspect ratio, small feature etch without significant mask erosion. Alternatively, it may be possible to take a

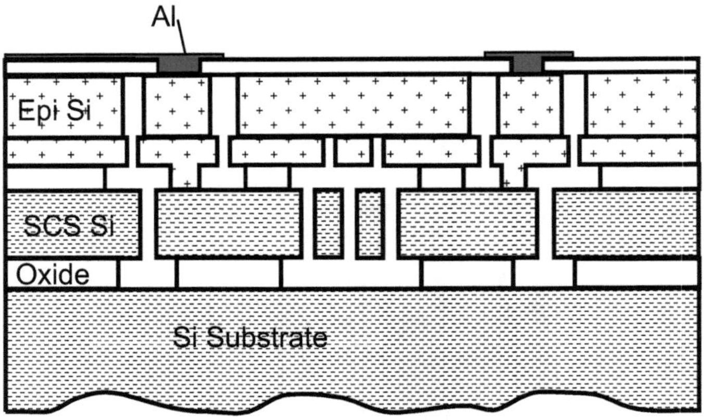

Fig. 5.6 [19] Schematic of the process cross-section of an encapsulated beam resonator using the EpiSeal^{TM} process (©SEMI 2005), reprinted with permission.

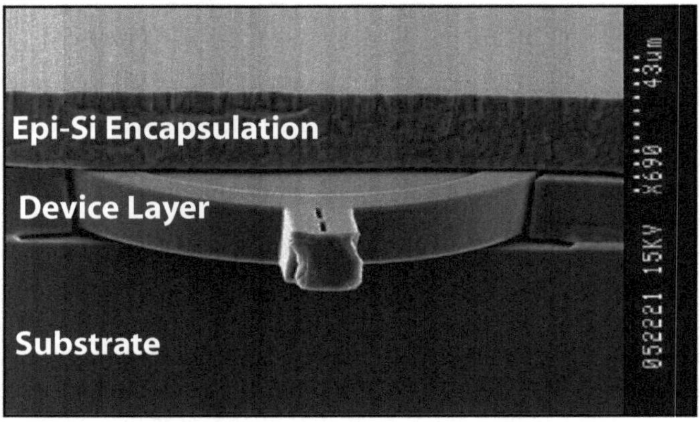

Fig. 5.7 [19] SEM image of the cross-section of an encapsulated resonator (©SEMI 2005), reprinted with permission.

lesson from the transferred cap ribbing approach. Depositing a very thick sacrificial oxide allows using a two stage etch to create a mold that will define a ribbed scaffold membrane if refilled with a conformal SiC film (Figure 5.8) [25]. This reduces the constraint on the scaffold deposition process to produce very thick, low-stress films in order to achieve the necessary mechanical rigidity. It also reduces the aspect ratio needed for etch access holes, reducing the constraint on the etch process. Finally, the etch access holes can be sealed using PECVD SiC [27] or IBAD SiC in order to prevent mass loading of the encapsulated resonator structures [26]. In this approach it is not clear that an epitaxial SiC layer of sufficient planarity and necessary quality for post-MEMS IC integration can be grown on top of this type of arrangement of films since epitaxial growth tends to mimic the crystallinity of the underlying layer. Fortunately though, another avenue might be open to those developing SiC-based

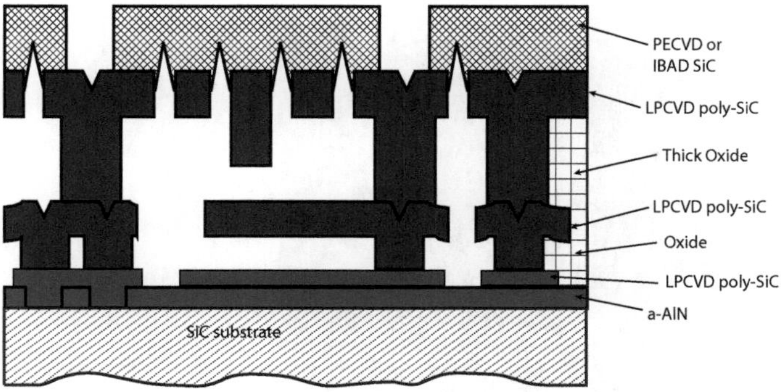

Fig. 5.8 Schematic cross-section of a process to use a thick sacrificial scaffold oxide to create a ribbed membrane from a thinner SiC film to enhance the mechanical robustness of the resulting encapsulation membrane.

microsystems. Since the thermal budget limits for SiC are considerably higher than silicon, it may be possible to instead attempt a post-IC MEMS integration. The processes outlined above are below 800 °C and might be able to be pushed somewhat lower, which may be sufficient for full MEMS and IC integration.

Another variation of the above silicon processes has been recently explored, which sandwiches the device layer in a two-device-layer SOI stack (Figure 5.9) [28]. Thus, the scaffold layer is made using a wafer bonding technique (Section 5.1.1.1) but the sealing layer is still an epitaxial silicon deposition process. This removes the constraint that the scaffold sacrificial layer actually refill all the device layer trenches. This constraint limits device topology to structures that do not require large gaps to allow for large proof mass motion [29] or for creating offset gap capacitive array position sensing [30]. This type of approach may be feasible for SiC as well, and would be a likely candidate for enabling post-MEMS IC integration. But this remains an open area of research for SiC technology.

As has been outlined, some successful silicon deposition based packaging concepts are strong contenders for transition to SiC materials for 0-level packaging. The authors believe ultimately this will be the direction that high-temperature, corrosion resistant sensor packaging will ultimately pursue; however, several harsh environment applications do not require high temperature operation (Chapter 1) but are limited by corrosion survival. Because of the relative processing ease of the coating approach, SiC coating of silicon-based technology has been the dominant SiC-based packaging technique demonstrated to date.

5.1.2.2 SiC Chemical Passivation Coating

A subset of packaging uses SiC coating as a protectant for more vulnerable materials such as silicon. These thin coatings, typically over silicon, impart a chemical resistive shell over the structural component. Since the coatings are not meant to be structural themselves, these coatings might only be tens to hundreds of nanometers

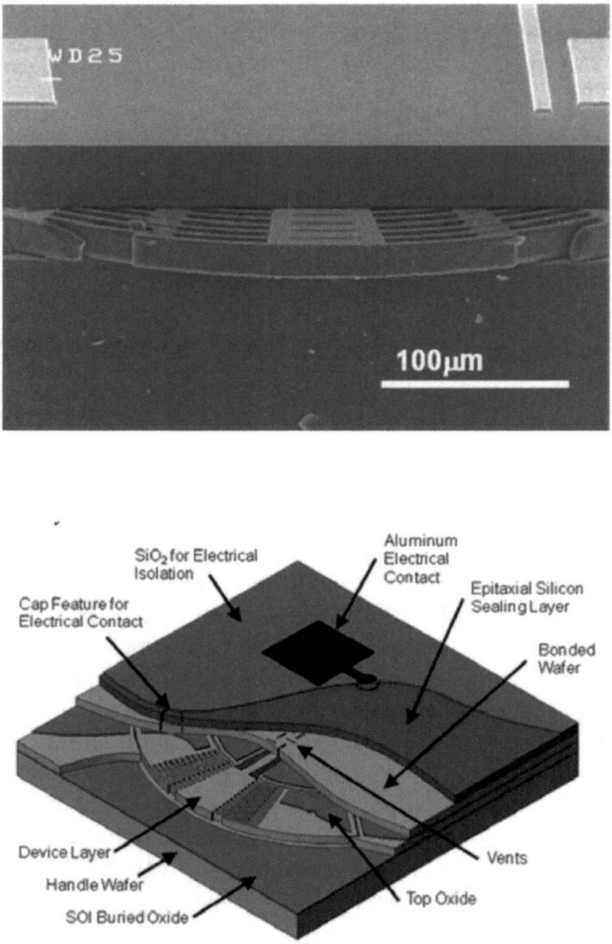

Fig. 5.9 [28] (a) Schematic cross-section of bonded scaffold layer and deposited sealing layer alternative to the EpiSeal^{TM} and (b) SEM image of the resonant proof-mass structure (©TRF 2010), reprinted with permission.

thick — enough to ensure the coating does not contain pinholes. LPCVD SiC deposition may be used, especially if a highly conformal film is desired. PECVD SiC has also been successfully used and is deposited at significantly lower temperatures.

The higher deposition temperature of LPCVD provides energy to allow migration of the depositing molecules to move along the deposition surface. This tends to lead to a conformal film since the molecules do not get locked locally due to the energy barrier of migration but can move to a low thickness region, which happens

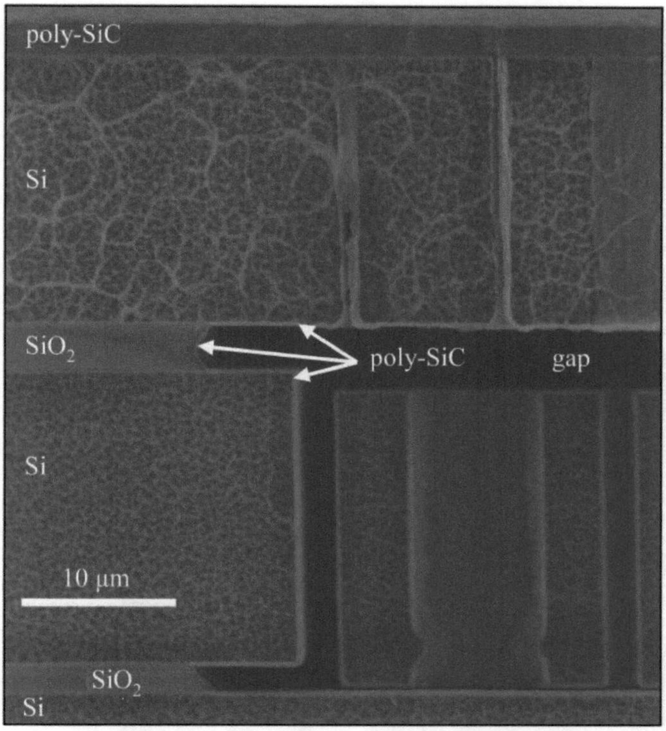

Fig. 5.10 [32] SEM image of the resonant structure, showing the relative thickness of the SiC coating on the devices below the scaffold layer (©IEEE 2009), reprinted with permission.

to be a lower energy state. This technique has already been discussed in the context of coating unencapsulated micro-structures in Section 4.2.1, demonstrating that corrosion protection is possible with tens of nanometers of LPCVD SiC films.

LPCVD SiC has also been used as the sealing deposition layer for the epitaxial silicon 0-level packaging discussed in Section 5.1.2.1. In this case, a thick LPCVD film of approximately 2.1 μm was deposited at a temperature of 800 °C using a 1,3-disilabutane (DSB) precursor [31]. Because of the relative conformality of this deposition, the resonator structures were also coated with approximately 300 nm of SiC (Figure 5.10). This results in an encapsulated structure that is resistant to chemical etching through the scaffold material and can act as a non-stiction coating for the released structures; however, inconsistent mass loading of the resonators across a wafer will need to be addressed. PECVD or IBAD a-SiC could likewise be used as the sealing layer if sufficient thicknesses can be achieved. This would reduce concern for mass-loading of the encapsulated devices. IBAD a-SiC is currently limited to about 1 μm thickness because of intrinsic film stress [26].

Another interesting scaffolding technique is available for silicon microsensor packaging. It has been successfully used for creating pressure sensors [33], fluidic microchannels [34], and encapsulated resonators [35]. By tuning the LPCVD

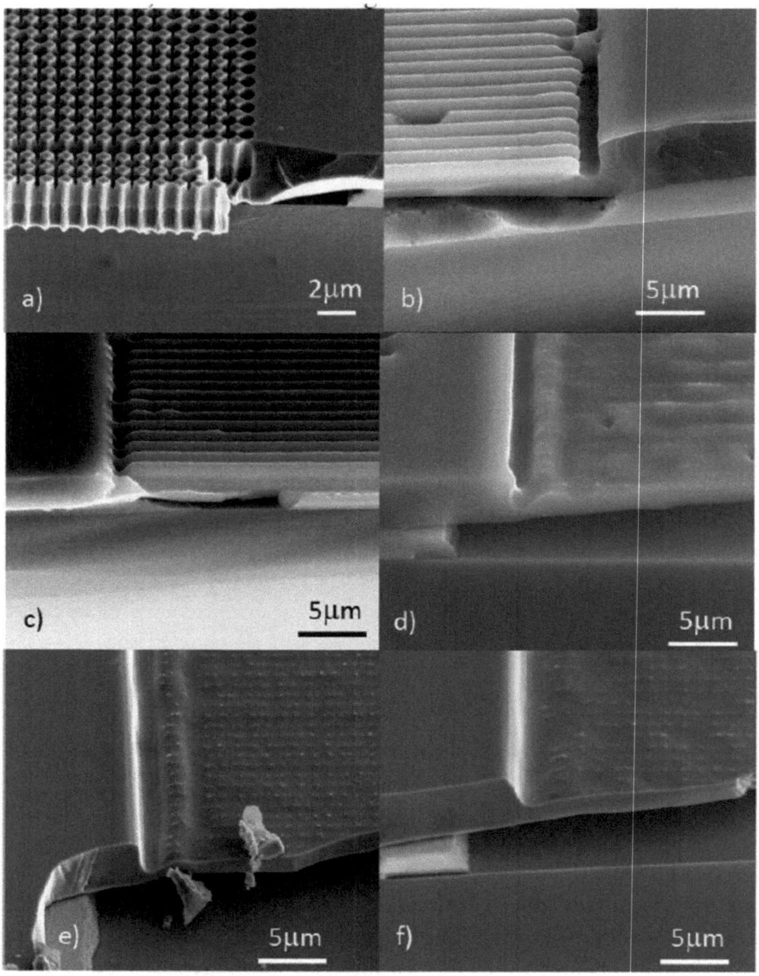

Fig. 5.11 [37] SEM image of silicon migration in a silicon scaffold layer at 1100 °C. (a)-(f) represent different times at temperature with (a) being the initial unprocessed state (©TRF 2010), reprinted with permission.

polysilicon deposition process [36] or by electrochemical etching of the device layer of an SOI wafer, it is possible to create a porous silicon membrane that allows etchant to penetrate through the porous layer to create a free-standing membrane (Figure 5.11). After the sacrificial layer is etched away using either a wet or vapor etch, the scaffold is sealed during the subsequent deposition step. Because the voids in the silicon are extremely small and not necessarily directional, rapid sealing of the membrane occurs with little deposition into the cavity. Thus, limitation in thickness of currently PECVD or IBAD a-SiC processes is not a barrier in providing a corrosion resistant coating if a porous silicon scaffold layer is used.

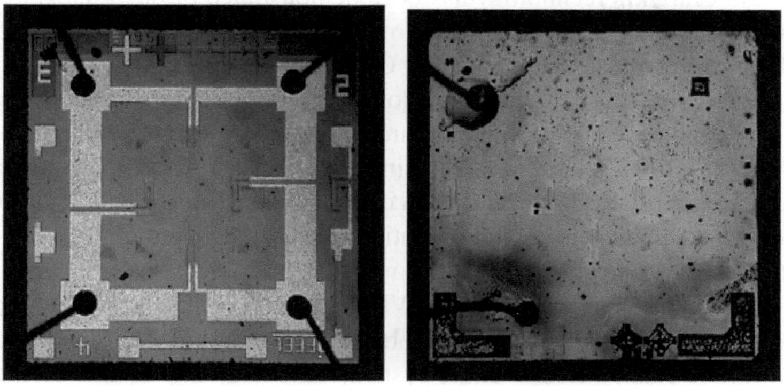

(a) (b)

Fig. 5.12 [39] Micrographs of (a) PECVD coated and (b) uncoated silicon pressure sensor after 20m and 1m in KOH, respectively (©IOP 2007), reprinted with permission.

There is a variation on the porous scaffold layer technique available in silicon. By making the release holes narrow enough or by utilizing a porous silicon layer, silicon migration through high-temperature treatment can be used to seal off the scaffold layer. There is a volume conservation thinning of the resulting membrane and the process temperatures are quite high, which prevent depositing silicon electronics first [37, 38]. This sealing approach likewise removes the constraint on SiC thickness to refill the etch access holes since they are self-sealed; only enough a-SiC thickness to provide the required corrosion protection is needed.

PECVD a-SiC is being actively explored for protection of silicon pressure sensors as well. For instance, a 1 μm thick PECVD a-SiC film was applied to a commercial pressure sensor (First MEMS, Co. Model YX-PS500A-150E) [39]. The coating was shown to protect the silicon membrane and exposed Al traces up to 20 minutes in KOH, whereas the uncoated device was destroyed in under 5 minutes (Figure 5.12). The pressure sensor was characterized with the coating. It exhibits approximately 11 % lower sensitivity due to the increased membrane stiffness; however, it also exhibited a reduced temperature sensitivity, presumably due to the difference in CTE between the silicon membrane and SiC coating, which would tend to increase membrane stress with increased temperature, counteracting the decrease in elastic modulus of the membrane with increased temperature.

Another interesting application for SiC coatings is in the field of implantable electronics and sensors for medical applications. The human body actively attacks foreign objects, hence a high degree of corrosion resistance is necessary. Implanted silicon and silicon dioxide films will etch over time. Hence, long term applications require a corrosion-resistant coating. This coating must also be non-toxic. Although not alone, SiC is a coating solution that can withstand long-term exposure to the body [40, 41].

Proteus Biomedical, Inc. has developed an encapsulation approach, termed ChipSkin^{TM}, to embed silicon CMOS circuitry into a corrosion resistant package.

Validation of the ChipSkin technology has been demonstrated through integration with an implanted cardiac pacing lead to improve the directionality and overall configuration of the pacing pulse. By integrating CMOS in the lead, the directionality of the pacing pulse becomes configurable, allowing efficient capture of heart muscle tissue at low voltages while preventing phrenic nerve capture at even high pacing voltages [42]. This is accomplished by using a multi-layered coating approach, which includes SiC, to protect both the CMOS circuitry and electrical interconnects. This approach has exhibited successful operation for the equivalent of 50 years of pacing *in vitro* and over one year of pacing *in vivo*.

Very low temperature a-SiC_x:H using PECVD has been investigated as a biocompatible coating for a wireless silicon neural probe array system [43]. This application requires a protective coating for the entire system, including the polymeric underfill in the wireless device, which restricted deposition temperature to below 200 °C. Even at these low temperatures, Si-C bond density control was possible by varying the hydrogen dilution ratio and precursor ratio. Optimization at 200 °C showed using a hydrogen dilution ratio of 5.71 with a precursor ratio of 0.17 ("saline starving" condition), resulted in reduced film stress and good Si-C bond density. Conformality was difficult to achieve over the full 1.2 mm length of the probe tips, ranging from 0.55 μm at the base to 2.5 μm at the tip. 650 $\AA$ thick films demonstrated no detectable dissolution or defects of the film after 6 weeks passive soak in phosphate buffered saline at 90 °C. Furthermore, the films maintained a high impedance after 6 months in Ringer's solution at 37 °C based on spectroscopy measurements.

5.2 First-level MEMS Packaging

First-level packaging typically completes the zero-level packaging such that the IC circuitry or MEMS device is encased in additional packaging to create a module that has the necessary interconnects to interface with a printed circuit board either by direct soldering or insertion into a socket (Figure 5.2). Traditionally, this would include metal leads or array of pins embedded in a polymer mold to create a lead frame. The sensor or electronics substrate is aligned and bonded to this lead frame using a low temperature adhesive to enable wirebonding or bump-bonding of the zero-level package interconnects to the lead frame traces. The package is then overmolded with an epoxy or resin to completely cover the zero-level package and bond wires or bump bonds, leaving only the desired lead frame tips or pins exposed. The overmold and lead frame mold material can be replaced with low-temperature co-fired ceramics (LTCC) or even high-temperature co-fired ceramics (HTCC), depending on the temperature stability required of these components.

Military-grade hermetic packaging though is still generally the welded metal can with glass frit sealed electrical feedthroughs. From a corrosion resistance perspective, a stainless steel package is generally adequate. By providing a via through the metal can allows the packaged sensor to interact with the environment, as in pressure or chemical sensing. When the package must be compromised in this way, the zero-level packaging must be equally corrosion resistant. Also, the bond wires or bump bonds must either be sealed internally within the package to prevent exposure to the environment, be coated or embedded in a corrosion resistant layer, or be themselves corrosion resistant. Consistent nickel and platinum wirebonding is an area of research for this reason, although depending on the environment and operating temperature, gold wirebonds may be sufficient and use a well developed process. Developments in ceramic packaging, thick film metallization, and electrical interconnects will be discussed in greater detail.

5.2.1 Thermal Expansion Matching of First-level Packaging Materials

For very high temperature applications, even metal packaging causes issues of thermo-mechanical stress of the die and wirebonds. Therefore, using materials that are well matched to SiC need to be considered. Aluminum oxide (alumina), aluminum nitride, and beryllium oxide are ceramics that show promise for corrosion resistant, high temperature first-level packaging of SiC devices [44]. Alumina and aluminum nitride are of particular interest not only because they exhibit good thermal conductivity and are well matched to SiC in terms of CTE but also because there is considerable knowledge in how to cast or machine components from these materials. Alumina in particular is often used in etch chamber ceramics for instance, which can require fabrication of complex, delicate shapes; however, the CTE of alumina differs more from SiC at high temperatures than aluminum nitride, making alumina less favorable (Figure 5.13). Although not found in the microsystem literature, sintered a-SiC is another possible first-level packaging material. It is not as commonplace as alumina, but is now being used in non-microsystem applications such as low-friction surfaces for bearings [45] and specialty optics [46]. Sintered a-SiC is used commercially for structural applications in sheet or pre-formed shapes and based on the literature has a closely matched thermal expansion rate to single-crystalline SiC substrates [47]. Hence, this is another material worth investigating for first-level packaging.

Thermal simulations of packaging for SiC substrates were carried out by L-Y Chen *et al.* These simulations confirm that the use of ceramics with close CTE to SiC reduces internal stress levels compared to standard packaging material for operating at 500 °C. The use of a thick gold adhesive layer for die attach was also simulated, indicating that the soft characteristics of the gold die attach is beneficial in reducing stress on the die by its ability to yield before transmitting high strain levels to the die [49]. Thick Au die attach along with aluminum nitride packaging can reduce the stress level in the die by a factor of 0.75 compared to traditional packaging. Although patterned bond pads were not explored, this is another avenue of providing stress-relief in the die attach for high temperature operation.

Thick gold or thick platinum metallization is used in conjunction with these ceramic systems and has been used to create both first-level packaging as well as basic

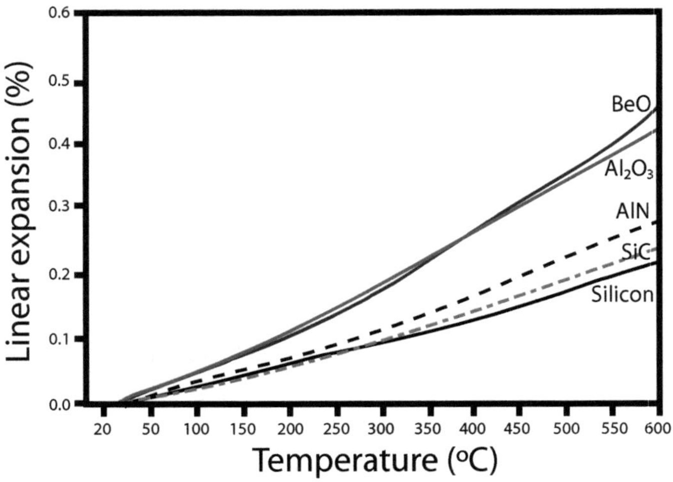

Fig. 5.13 [44] Linear expansion rates of select ceramics compared to silicon and silicon carbide from room temperature to 600 °C (©IEEE 2002), reprinted with permission. Note the slope of these curves represents the CTE.

circuit boards (Figure 5.14) [48]. Aluminum nitride sheet shaped using electrodischarge machining (EDM) has also been used as the first-level packaging covers for SiC pressure sensors [44]. A Kovar header was also used as the mechanical interface and stress relief to a stainless steel threaded housing for sensor mounting because Kovar is a nickel iron alloy that is CTE matched to silicon, decreasing the CTE mismatch being built into the overall system (Figure 5.15). The packaged electronics and sensor have been successfully operated up to 600 °C in oxidizing environments, demonstrating that careful design using CTE matched materials can be used to create first-level packaging even for applications requiring greater than 500 °C operation.

This deals with the primary concern of first-level packaging: namely the encapsulation. Electrical interconnects pose a similar concern, although some interesting solutions have been investigated to date.

5.2.2 Electrical Interconnects

As was discussed in Chapter 3, corrosion resistant metals that can make good contact to SiC are being readily explored. But these initial contact metals need to continue to be routed out through the packaging layers. Standard gold bond wires begin to degrade at prolonged exposure to a 500 °C oxidizing environment, exhibiting a degradation in wire resistance with time. Replacing the thin (25 μm diameter) wire with a thicker gold ribbon allowed satisfactory performance under the same conditions. When paired with thick gold metallization onto the ceramic first-level

Fig. 5.14 [48] Image of die level packages of SiC IC components put together on a custom alumina IC printed circuit board for testing electronics performance up to 500 °C (©SPIE 2006), reprinted with permission.

packaging, interface stresses between the wirebond tip and board-level metallization is likely to be sufficient, even at high operating temperatures; however, simulation of these types of conditions is likely needed to gain better insight. This may also push further constraints onto the electrical contact metal stack, requiring careful matching of the final contact metal to the ribbon bonding used.

Platinum wirebonding is also of interest because of its high corrosion and temperature resistance; however, it is a difficult material to bond using traditional ultrasonic bonding. Temperature-assisted ultrasonic bonding, as is used for gold wirebonding, is one approach, although research in this area is not well explored. A group from General Motors has shown platinum wire wedge bonding can be used and survives operations as high as 900 °C for one hour without degraded electrical performance [50].

These first-level packaging concerns highlight that additional packaging and requisite interconnects are a reliability concern for operating at high temperature or in corrosive media. Although this is also true in the general microsystem sense, reducing interconnect points and layers is expected to ultimately improve the overall system reliability. This can be achieved by including as much as function in the sensor substrate as possible through monolithic integration, even though this increases the manufacturing difficulty. Because temperature and corrosion exacerbate these failure mechanisms, mitigating them as much as possible through monolithic integration becomes especially important for harsh environment microsystems.

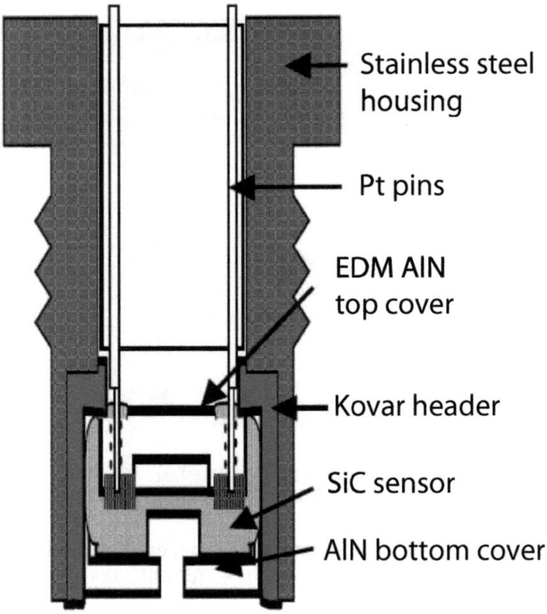

Fig. 5.15 [44] Schematic cross-section of a multi-material pressure sensor housing specifically designed for operation at 500 °C (©IEEE 2002), reprinted with permission.

References

1. Tummala RR, Rynaszewski EJ (1997). Microelectronics Packaging Handbook, Part II: Semiconductor Packaging. Springer. ISBN: 0412084317
2. Hoya Corporation, Optics Division, corporate website: www.hoyaoptics.com
3. Tong QY, Lee TH, Werner P, Gosele U (1997). Fabrication of Single Crystalline SiC Layer on High Temperature Glass. J. Electrochem. Soc. 144:L111–L113
4. Tudryn CD (2004). Characterization of Anodic Bonding. Masters Thesis, Department of Mechanical Engineering and Material Science, Massachusetts Institute of Technology.
5. Di Cioccio L, Tiec YL, Letertre F, Jaussaud C, Bruel M (1996). Silicon Carbide on insulator formation using the Smart Cut Process. Electronics Letters. 32:1144–1145.
6. Sparks D, Queen G, Weston R, Woodward G, Putty M, Jordan L, Zarabadi S, Jayakar K (2001). Wafer-to-wafer bonding of nonplanarized MEMS surfaces using solder. J. Micromechanics and Microengineering 11(6):630–634
7. Shoaf SE, Feinerman AD (1994). Aligned Au-Si eutectic bonding of silicon structures. J. Vac. Sci. Technol. 12:19–23
8. Cheng Y, Hsu W, Najafi K, Nguyen C-T, Lin L (2002). Vacuum packaging technology using localized aluminum/silicon-to-glass bonding. JMEMS 11:556–565
9. Kim S, Seo YH, Cho Y, Kim GH, Bu JU (2003). Fabrication and characterization of a low-temperature hermetic MEMS package bonded by a closed loop AuSn solder-line. Proceedings IEEE Sixteenth Annual International Conference on Micro Electro Mechanical Systems, Jan. 19-23, 2003:614–617
10. Baggerman AFJ, Schwarzbach D (1998). Solder-jetted eutectic PbSn bumps for flip-chip. IEEE Transactions on Components, Packaging, and Manufacturing Technology Part B: Advanced Packaging, 21:371–381

11. Cheng YT, Lin L, Najafi K (1999). Localized bonding with PSG or indium solder as intermediate layer. Proceedings of the 12th IEEE Int. Conf. MEMS, Jan 17-21, 1999:285–289
12. Maharbiz MM, Cohn MB, Howe RT, Horowitz R, Pisano AP (1999). Batch micropackaging by compression-bonded wafer-wafer transfer. 12th IEEE Int. Conf. MEMS, Jan 17-21, 1999:482–489
13. Heck JM (2001). Polycrystalline silicon germanium for fabrication, release, and packaging of microelectromechanical systems. Ph.D Thesis, Applied Science and Technology, University of California, Berkeley.
14. Lin L, Howe RT, Pisano AP (1998). Microelectromechanical filters for signal processing. JMEMS 7:286–294
15. Stark BH, Najafi K (2004). A low-temperature thin-film electroplated metal vacuum package. JMEMS 13:147–157
16. Candler RN, Park WT, Li HM, Yama G, Partridge A, Lutz M, Kenny TM (2003). Single Wafer Encapsulation of MEMS Devices. IEEE Transactions on Advanced Packaging 26(3):227–232
17. Höchst A, Scheuerer R, Stahl H, Fischer F, Metzger L, Reichenbach R, Lärmer F, Krönmuller S, Watcham S, Rusu C, Witvrouw A, Gunn R (2004). Stable thin film encapsulation of acceleration sensors using polycrystalline silicon as sacrificial and encapsulation layer. Sensors and Actuators A 114(2-3):355-361
18. Monajemi P, Joseph P, Kohl PA, Ayazi F (2006). Characterization of a Polymer-Based MEMS Packaging Technique. Proc. IEEE Advanced Packaging Materials, Atlanta, GA, Mar. 2006:139–144
19. Partridge A, Lutz M, Kim B, Hopcroft M, Candler RN, Kenny TW, Petersen K, Esashi M (2010). MEMS Resonators: Getting the Packaging Right. Semicon Japan 2005.
20. Azevedo RG, Jones DG, Jog AV, Jamshidi B, Myers DR, Chen L, Fu, XA, Mehregany, M, Wijesundara, MBJ, Pisano, AP (2007). A SiC MEMS Resonant Strain Sensor for Harsh Environment Applications, IEEE Sensors, 7(4):568–576
21. Gao D, Wijesundara MBJ, Carraro C, Howe RT, Maboudian R (2004). Transformer coupled plasma etching of 3C-SiC films using fluorinatedchemistry for microelectromechanical systems applications. Journal of Vacuum Science Technology B 22(2):513–518
22. Tanaka S, Rajanna K, Abe T, Esashi M (2001). Deep Reactive Ion Etching of Silicon Carbide. J. Vac. Sci. Technol. B 19:2173–2177
23. Lazar M, Vang H, Brosselard P, Raynaud C, Cremillieu P, Leclercq J-L, Descamps A, Scharnholz S, Planson D (2006). Deep SiC etching with RIE, Superlattices and Microstructures, E-MRS 2006 Symposium S 40(4-6):388–392
24. Jones DG, Pisano AP (2010). Aluminum nitride as a masking material for the plasma etching of silicon carbide structures. IEEE 23rd International Conference on Micro Electro Mechanical Systems, Hong Kong, Jan. 24-28:352–355
25. Azevedo RG (2007). Silicon Carbide Micro-extensometers for Harsh Environments. Dissertation, Department of Mechanical Engineering, University of California, Berkeley.
26. Jones DG, Azevedo RG, Chan M, Pisano AP, Wijesundara MBJ (2007). Low temperature ion beam sputter deposition of amorphous silicon carbide for wafer-level encapsulation. MEMS 2007, 20th IEEE SENSORS 2007 Conference on Micro and Nano sensors Technical Digest 2007:275–278
27. Rajaraman V, Pakula LS, Pham HTM, Sarro PM, French PJ (2008). Application of PECVD a-SiC Thin-Film Layer for Encapsulation of Microstructures. 11th Annual Workshop on Semiconductor Advances for Future Electronics and Sensors:609–612
28. Messana MW, Graham AB, Yoneoka S, Howe RT, Kenny TW (2010). Packaging of Large Lateral Deflection MEMS Using a Combination of Fusion Bonding and Epitaxial Reactor Sealing. Solid-State Sensor, Actuator and Microsystems Workshop, Hilton Head Island, South Carolina, June 6-10:336–339
29. Rajgopal S, Zula D, Garverick S, Mehregany M (2009). A Silicon Carbide Accelerometer for Extreme Environment Applications. Materials Science Forum 600-603:859–862
30. Yeh R (2001). Articulated mechanisms and electrostatic actuators for autonomous microrobots. Ph.D dissertation, University of California, Berkeley, Deparment of Electrical Engineering

31. Yoneoka S, Roper CS, Candler RN, Chandorkar SA, Graham AB, Provine J, Maboudian R, Howe RT, Kenny TW (2010). Characterization of encapsulated micromechanical resonators sealed and coated with polycrystalline SiC. JMEMS 19(2):357–366
32. Roper CS, Candler R, Yoneoka S, Kenny T, Howe RT, Maboudian R (2009). Simultaneous wafer-scale vacuum encapsulation and microstructure cladding with LPCVD polycrystalline 3C-SiC. Solid-State Sensors, Actuators and Microsystems Conference, Denver, CO, June 21-25, 2009:1031-1034
33. Marek J (2007). MEMS Technology – From Automotive to Consumer. MEMS 2007, 20th IEEE SENSORS 2007 Conference on Micro and Nano sensors Technical Digest 2007:59–60
34. Dougherty GM, Sands T, Pisano AP (2003). Microfabrication using one-step LPCVD porous polysilicon films. JMEMS 12(4):418–424
35. He R, Fan L, Wu MC, Kim C-J (2004). Porous Polysilicon Shell Formed by Electrochemical Etching for On-Chip Vacuum Encapsulation. Solid-State Sensor, Actuator and Microsystems Workshop, Hilton Head Island, South Carolina, June 6-10:332–335
36. Leibouitz KS, Pisano AP, Howe RT (1995). Permeable polysilicon etch-access windows for microshell fabrication. 8th Int. Conf. on Solid-State Sensors and Actuators, Stockholm, Sweden, June 1995, 1:224–227
37. Provine J, Ferralis N, Graham AB, Messana MW, Kant R, Maboudian R, Kenny TW, Howe RT (2010). Time Evolution of Released Hole Arrays into Membranes Via Vacuum Silicon Migration. Solid-State Sensor, Actuator and Microsystems Workshop, Hilton Head Island, South Carolina, June 6-10:344–347
38. Kant R, Choo H (2010). Numerical Modeling and Experimental Verifications of Single-Step, Deposition-Free, Hermetic Sealing Using Silicon Migration. Solid-State Sensor, Actuator and Microsystems Workshop, Hilton Head Island, South Carolina, June 6-10:262–263
39. Zhang H, Guo H, Wang Y, Zhang G, Zhihong L (2007). Study on a PECVD SiC-coated pressure sensor. J. Micromech. Microeng. 17:426–431
40. Kotzara G, Freasa M, Abelb P, Fleischman A, Roy S, Zorman C, Morane JM, Melzak J (2002). Evaluation of MEMS materials of construction for implantable medical devices. Biomaterials 23:2737–2750
41. Zorman C (2009). Silicon Carbide as a Material for Biomedical Microsystems. DTIP of MEMS & MOEMS, Rome, Italy, Apr. 1-3, 2009:hal-00395712
42. Azevedo R, Costello B, Frank J, Jensen M, Thompson T, Zdeblick M (2010). Novel Method of Protecting and Connecting CMOS Chips Enables Networked Pacing Leads. Solid-State Sensor, Actuator and Microsystems Workshop, Hilton Head Island, South Carolina, June 6-10:260–261
43. Hsua J-M, Tathireddyb P,Rietha L, Normannc AR, Solzbacher F (2007). Characterization of a-SiCx:H thin films as an encapsulation material for integrated silicon based neural interface devices. Thin Solid Films 516(1):34–41
44. Savrun E (2002). Packaging Considerations for Very High Temperature Microsystems. Sensors Conference 2002:1139–1143
45. Saint-Gobain Ceramics sintered SiC product website: www.hexoloy.com
46. CoorsTek Advanced Ceramics corporate website: www.coorstek.com
47. Ganesh I, Jana DC, shamshad S, Thiyagarajan N (2006). An Aqueous Gelcasting Process for Sintered Silicon Carbide Ceramics. J. American Ceramics Society 89:3056–3064
48. Hunter GW, Wrbanek JD, Okojie RS, Neudeck PG, Fralick GC, Chen LY, Xu J, Beheim GM (2006). Development and application of high temperature sensors and electronics for propulsion applications. Proceedings of the SPIE Defense and Security Symposium, Sensors for Propulsion Measurement Applications Workshop, 2006.
49. Chen L-Y, Lei J-H (2006). Packaging of Harsh Environment MEMS Devices. Chapter 12 of The MEMS Handbook, Gad-el-Hak M editor. CRC Press, Boca Raton.
50. Mantese JV, Alcini WV (1988). Platinum Wire Wedge Bonding: A New IC and Microsensor Interconnect. Journal of Electronic Materials 17(4):285–289

Chapter 6
System Integration

A microsystem is a collection of integrated devices that contains MEMS (sensors, actuators, and timing devices), electronics (control, sense, and data processing), communication (wired or wireless), and a power source. Figure 6.1 schematically illustrates a complete autonomous microsystem. Realization of all these components into a single system is rather complex. Several integration approaches have been used or proposed for conventional microsystems. Application requirements, performance advantages, manufacturability, and cost advantages drive which integration route is ultimately used.

Depending on the required performance specifications, device characteristics, and restrictions imposed by the application environment, the microsystem complexity can vary from MEMS with simple impedance matching buffer electronics to MEMS with drive, sense, and signal processing electronics to fully autonomous systems of MEMS with drive, sense, and signal processing electronics integrated with wireless communication and power. In most cases, microsystem integration refers to integration of MEMS with drive, sense, and signal processing electronics. The power and the communication are considered as subsystems or modules. This approach reduces the complexity of integration schemes and process incompatibilities can be avoided. In this chapter, SiC electronics to MEMS integration schemes will be detailed separately from communication and power, which will be discussed in terms of available technologies, ability of integration into the microsystem, and suitability for harsh environment applications.

6.1 SiC MEMS and SiC Electronics Integration

SiC MEMS and SiC IC integration is still at an infancy stage, and no integration efforts have been reported in the literature. This is mainly because both SiC MEMS and electronics have not reached the same maturity level as Si technology. Borrowing from MEMS-IC integration in Si technology, integration schemes for SiC can be proposed. Silicon technology uses two main strategies for MEMS-IC integration,

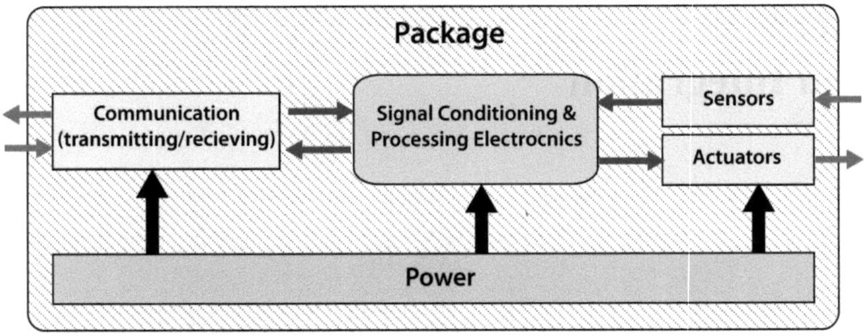

Fig. 6.1 Conceptual drawing of fully integrated autonomous microsystem.

a hybrid integration which involves two or more chips and monolithic integration in which MEMS and IC are fabricated side by side or on top of each other. Similar schemes can be exploited for SiC technology as well.

In the case of hybrid integration, MEMS and ICs are fabricated separately and integration is done through wire bonding or flip-chip bonding before packaging them together. In some cases, separately packaged MEMS and ICs are integrated at the board level. Monolithic integration uses concurrent fabrication to realize both electronics and MEMS on the same chip. Realization of the integrated system needs both MEMS and electronics to be designed to common system specifications and the interface complexities, performance specifications, and process-related issues should be addressed at the design level. At the current maturity level of the technology, SiC MEMS and electronics development are not integrated. Each one is confined to its own development path since there are many device level issues including dielectric stability, interconnect reliability, and materials quality that need attention. To date, SiC MEMS is centered around individual device level demonstration while SiC electronics efforts are mainly focused on high power and high temperature electronics. Little effort has been reported on signal conditioning electronics for MEMS. A recent report by Patil *et al.* shows development of SiC JFET based ICs targeted for microsensor applications[1]. This is a very good initial step towards developing SiC integrated circuits for MEMS based sensors operating in harsh environments; however, actual integration has not yet been reported. Potential techniques for MEMS and IC integration will be discussed in this chapter.

6.1.1 Multi-chip Hybrid Approach

In the hybrid approach, each component or subsystem is fabricated separately. Since process and device performance optimization is done independently from each other, this approach is advantageous in terms of the development time of each subsystem. The development of MEMS and electronics independently from one another also decouples the thermal budget and process constraints that are common to integrated fabrication. Furthermore, this approach may enable cost reduction as cost and yield optimization can be done for each individual subsystem separately. However, the system integration aspects with regards to interface physics as well as compatibility with assembly and packaging must be taken into consideration when MEMS and electronics are designed.

SiC technology, both MEMS and electronics, is relatively young and immature. With regards to electronics, individual devices have been demonstrated or are in use in many application environments. SiC ICs are still at an early research stage and tremendous amount of research and development are needed before they are a viable technology for harsh environment microsystems. Despite the late start, SiC MEMS are at relatively advanced stage, and some are ready for real-world applications. Nonetheless, there still remain many advances necessary at the component level before considering the monolithic approach. Although power and switching applications have been demonstrated, SiC electronics compatible with MEMS operating requirements need to be developed. SiC MEMS technology also required further improvements from both materials and processing aspects. Therefore, the hybrid approach may be well suited for SiC microsystems at this stage as it will reduce the development time and can lead to system level demonstrations and some limited applications. Moreover, the hybrid approach allows the application of SiC MEMS for low temperature use, yet in the presence of other harsh conditions such as corrosive media, with the use of silicon electronics.

The hybrid approach is promising in the present context of SiC technology; however, in terms of performance as well as harsh environment compatibility, the hybrid approach has some inherent disadvantages. Performance-limiting disadvantages include parasitic capacitance and contact resistance of the interconnects, bond pads, and bond wires [2]. Furthermore, these types of connections are highly susceptible to electromagnetic interference and current leakage, which decreases signal strength. This can be a significant problem especially for thin-film, surface-micromachined structures in which position is sensed capacitively [2]. Some of the intended applications require these SiC microsystems to work in extremely high shock or vibration environments and high temperature conditions. The integrity of the interface would be an issue under these conditions whether the multi-chip approach is flip-chip, stacked chips, or board-level integration.

6.1.2 Monolithic Fabrication Approach

In the monolithic approach, both electronics and MEMS are co-fabricated on the same chip either side by side or one over the other. Incorporating all the device elements on a single chip provides distinct advantages over the hybrid integration approach. From the performance stand point, monolithic integration minimizes parasitic capacitance, contact resistance, electromagnetic interference, and current leakage. Therefore, monolithically-integrated systems allow measuring small signals in the presence of high levels of noise. This approach is particularly advantageous when integrating multiple sensors with electronics for measuring motion along different axes because this approach allows precise alignment of each sensor. With regard to packaging, MEMS device and electronics on the same chip reduces packaging complexity and improves reliability by minimizing the number of off-chip electrical connections [2]. Depending on the integration process complexity and yield, the cost per system may also be reduced significantly.

Monolithic integration of MEMS with ICs involves fabrication of disparate device types on the same substrate. This poses many constraints on every aspect of fabrication. The functionality at the device and system level as well as process compatibility issues need to be considered at the design stage. Designing processes that allow realization of a system with multiple device types potentially with packaging is a daunting task in monolithic integration. It is imperative to design process flows with each subsequent process step within the thermal budget of the previous layers. Furthermore, the underlying device layer should be protected chemically and electrically from subsequent process steps. All these considerations may lead to several design and process iterations before realizing a functional, integrated system, resulting in longer development time than the hybrid version. The unavailability of standard MEMS processes may further complicate the process development as each type of MEMS device may require process or design alterations. However, longer development times and the associated costs may be justified by performance and form factor requirements for specific applications. Particular to harsh environment applications, sensors and actuators in combustion environments, space environments, and high shock and vibration conditions require robust, small form factor microsystems.

SiC is well suited for monolithic integration in comparison to silicon. It has extremely high thermal stability from both an electrical and mechanical standpoint. Furthermore, impurity diffusion only happens above 1800 °C [5]. It is chemically inert and exhibits high radiation stability. All of these factors loosen the process constraints. Therefore, in theory, the co-fabrication of SiC electronics and MEMS on the same chip would be relatively straightforward compared to monolithic integration of Si based IC and Si MEMS. However, the current state of SiC electronics and MEMS may still need to be focused on further improvement with regards to processes and designs before monolithic integration can be realistically considered. As mentioned previously, more effort needs to be devoted to advancing SiC IC fabrication for MEMS applications. SiC MEMS processes are mostly confined to academic institutions with a few rare exceptions [3]. Research needs to focus on application specific SiC MEMS device fabrication and long-term reliability testing so that real world applications are clear. This enables the market to lead further development of this technology.

Although there are no integration schemes proposed or investigated to date, possible routes for integrated fabrication of SiC microsystems can be examined by considering the materials and process aspects as well as drawing parallels from well-developed silicon technology [4]. In silicon technology, monolithic integration of microsystems can be categorized into three general groups. Three integration schemes are namely post-CMOS MEMS (MEMS last) insertion, interleaved MEMS insertion, and pre-CMOS MEMS (MEMS first) insertion [6]. First, these three concepts will be briefly introduced before discussing the relevance of these methods with regards to SiC MEMS-IC integration.

Post-CMOS MEMS Insertion: Post-CMOS MEMS insertion is considered the ideal integration case. One advantage to this method is the ability of fabricating MEMS structures on top of the IC, which reduces the interconnect lengths and in the ideal case minimizes system size. In this scheme, the IC is fabricated first independent of the MEMS structures and zero-level packaging. Thus, separate foundries can utilize independently optimized processes. IC fabrication typically has much smaller dimensional tolerances needed from photolithography. Starting the IC process on the virgin polished substrate allows high fidelity pattern transfer leading to high device performance and yield. From the MEMS point of view, post-CMOS MEMS insertion is highly beneficial for smaller MEMS companies who may not have the capital to invest in a dedicated CMOS foundry, creating a high barrier to entry compared to other methods.

The most common issue with a post-CMOS approach is that in order to create the desired materials or modify their properties, the MEMS process inevitably requires deposition or annealing steps that exceed the thermal budget of the underlying IC components. For silicon ICs, these issues include junction spiking, metal hillocking, parameter shifts caused by dopant diffusion, and degradation of interconnects [2]. Thus, the maximum temperature allowed in the MEMS fabrication steps post-CMOS is in the range of 400-500 °C, which limits the usable material set and in turn achievable MEMS devices. In the case of SiC, dopant diffusion is not significant; however, degradation of metal contacts is an issue.

An example of a successful product that uses post-CMOS MEMS is the Texas Instruments digital micromirror display [7]. In this process, low temperature physical vapor deposition of metal structural layers is used to avoid thermal budget issues. Poly-SiGe, a material very similar to poly-Si in terms of strength, reliability, and quality factor, has also been investigated for post-CMOS MEMS insertion because poly-SiGe can be deposited at lower temperature [2]. Although this material shows promise, issues around stress and stress gradient as well as source gas cost need to be addressed for successful commercialization.

There are also other commercial successes that used post-CMOS processing. For instance, piezoresistive pressure sensors commercially available from Bosch GMBH [8] and Freescale, Inc. [9] are fabricated using bulk micromachining of standard CMOS chips. These schemes rely on the MEMS device layers to come from existing CMOS layers. Hence, the bulk Si substrate itself is preferred as the mechanical element since the IC device layers are rarely tuned for mechanical response by default.

Pre-CMOS MEMS Insertion: Pre-CMOS MEMS insertion maintains that separation of the processes is still possible but that high temperature steps can be used for MEMS fabrication because they occur first. This avoids compromising the MEMS structures by restricting choice of materials and processes because of thermal budget constraints. Post-MEMS processes typically begin by restoring planarization of the surface. The success of this planarization impacts subsequent lithographic resolution of the CMOS foundry. Even with this, care must be taken to bury potentially contaminating materials sufficiently that the wafers can be certified clean enough for introduction into CMOS foundry tools.

Often it is this latter issue that causes the most difficulty. Even running the wafer through typical MEMS tools is enough to contaminate the wafer from a CMOS foundry perspective because the front-end processes are extremely sensitive to even trace contamination. Additionally, a reverse thermal budget issue, although less severe generally, can occur if the initial high temperature CMOS processes cannot be leveraged for annealing the MEMS devices or if an even higher temperature anneal cannot stabilize the mechanical properties for the temperatures that will be experienced during the CMOS process [10, 11]. If this is the case, then stress control put in at the MEMS processing steps is compromised. Specialized high-temperature metals must be used if metals are part of the pre-CMOS MEMS process.

One common pre-CMOS integration scheme is to etch a trench below the surface of the wafer and then bury and seal the MEMS device prior to microelectronic device fabrication. After planarization and surface preparation, the wafer is used as the starting material for CMOS process. Once the CMOS is completed, the MEMS device is released. An impact of this particular integration method is that the CMOS and MEMS do not occupy the same location on the die, increasing the area of the chip. From a practical standpoint, state of the art electronics processing is not used because ultra-high planarity is not always achievable and, more importantly, too much cost is associated with the risk of contamination of the CMOS foundry tools, so when this is allowed it typically occurs at older fabrication lines where the infrastructure costs have long since been recuperated.

A slight modification to this fabrication process uses SOI wafers [12]. This SOI substrate process has been further improved and adapted by the industry [13]. Bosch has invested in a process that starts with an SOI wafer and patterns additional silicon dioxide and Si layers where the Si layers are formed using an epitaxial growth process [14]. The epitaxial growth process creates a thick, high quality Si layer that can be used for either MEMS device layers or even scaffolding layers for zero-level packaging (see Section 5.1.2). One disadvantage to this process is that it inherently limits the gaps that can be created in the device layer region to ensure planarity of the upper layers, which limits the design of MEMS devices that utilize large spanning structures; however, a work-around utilizing a bonded capping wafer is being developed that increases the flexibility of this process [15].

Interleaved MEMS Insertion (Frontend IC-MEMS-Backend IC/MEMS metallization): In practicality, most MEMS insertion methods are some degree of an interleaved approach. In a common realization of this approach, first, frontend transistors are defined. Then, the process is interrupted before creating backend interconnect metallization. Once MEMS structures are fabricated next to predefined IC, the final metallization for IC and MEMS-IC interface is performed [16]; however, interleaving the processes leads to design compromises that limit microsystem performance. For instance, poly Si MEMS typically require a high temperature (900 °C) stress relieving anneal, and the longer annealing time causes dopant diffusion in the transistor wells. This limits the thickness of the MEMS layer. Post-MEMS metallization lessen the thermal budget issues in comparison to post-CMOS MEMS insertion since metal migration is typically the most temperature sensitive reaction. In general though, both MEMS and electronics have to be co-fabricated in the same foundry in the interleaved approach, which requires partnering with a foundry willing to devote a portion of their development line to developing such an endeavor. Otherwise, the interleaved MEMS-IC integration occurs only at companies large enough to have their own CMOS line and can likewise drive development resources on that internal line to develop a non-standard CMOS process flow.

The interleaved approach for MEMS-IC integration is widely adopted in commercial production environments. Some of the systems fabricated using the interleaved approach include Analog Devices ADXL series accelerometers and ADXRS series gyroscopes [13], Infineon Technologies KP100 series pressure sensors [17], and Freescale MPXY8000 series pressure sensors [9].

The knowledge from silicon IC and MEMS integration concepts can be utilized to find potential paths for SiC MEMS integration with SiC electronics. Some of the possible scenarios are outlined in next few subsections.

6.1.2.1 Interleaved MEMS Insertion for SiC

The interleaved approach for SiC MEMS-IC integration is fairly similar to that of poly-Si MEMS and Si CMOS integration. A simplified version of a generic fabrication process flow is depicted in Figure 6.2. The process sequence starts with a SiC substrate that has all the necessary epitaxial layers for SiC ICs. As an example, the architecture for a SiC JFET is selected. Electronics fabrication starts with patterning the top p and n epi layers. This is followed by ion implantation to produce n^+ regions. At this point, the IC fabrication is interrupted and substrate is prepared for MEMS integration. An isolation dielectric layer between the IC and MEMS device is deposited on top, covering the entire wafer. The most suitable isolation dielectric layers for this process would be Si_xN_y or amorphous AlN as they closely match the CTE of SiC and have a comparable thermal stability. Furthermore, this isolation layer should have high selectivity to a variety of etch chemistries that are used to remove sacrificial materials common to SiC technology, namely SiO_2 or poly-Si, making Si_xN_y and amorphous SiC good choices. MEMS device fabrication starts with deposition of an electrical routing layer, which also serves as the electrical interface between the SiC IC and MEMS device. The most logical routing layer material would be poly-SiC as it would also be the structural material for MEMS. Other highly conductive and high temperature materials, such as heteroepitaxially grown SiC and TiN, can also be exploited for this purpose. After patterning the routing layer, a sacrificial layer, typically SiO_2, is deposited on the entire substrate surface. Then the sacrificial layer undergoes planarization and patterning. Planarization at this stage creates a more uniform thickness MEMS structural layer as well as allows finer geometrical feature resolution from the photolithographic patterning process, which allows narrow capacitor gap definition for instance. Next, the MEMS structural layer is deposited and patterned. At this stage, the MEMS process is stopped and back end metallization of both the electronics and MEMS device occurs. The

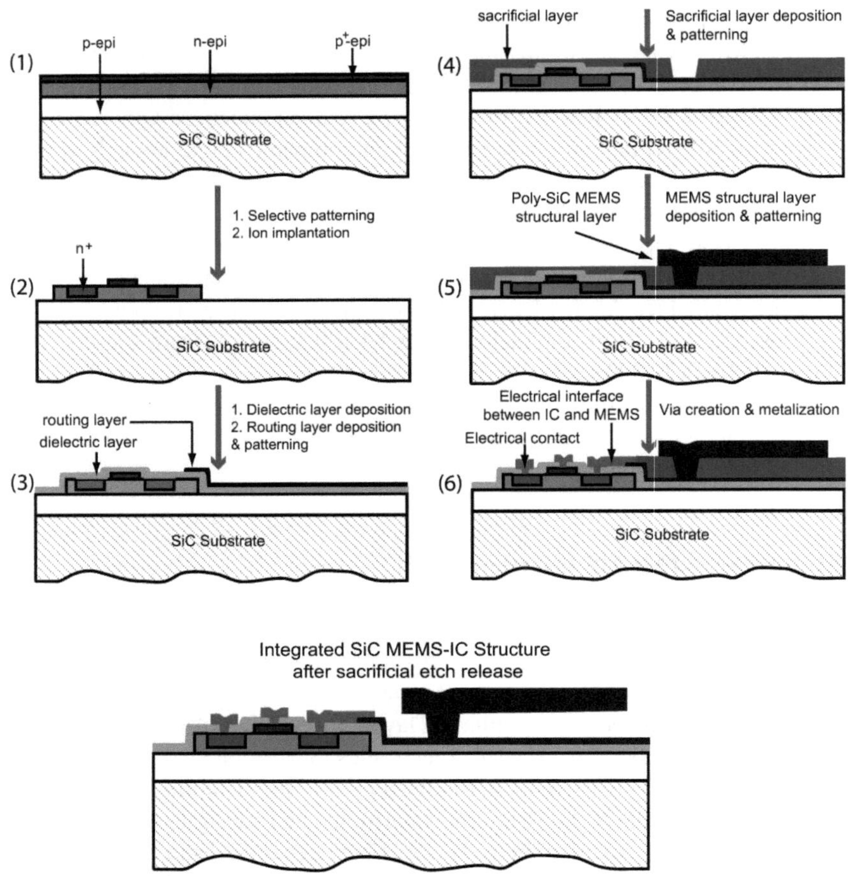

Fig. 6.2 A simplified version of interleaved SiC MEMS-IC integration.

MEMS-IC interface is created through this process. Finally, the sacrificial material will be removed with either a dry or wet etch process to create a released MEMS device integrated with electronics.

Typical poly-SiC MEMS processes are carried out at temperatures below 900 °C and this is 50% below the theoretical limits of doping impurity diffusion of SiC. Furthermore, there is no high temperature annealing process for stress relieving of poly-SiC as current LPCVD processes can produce low stress and low gradient films [18]. Therefore, many micron thick films are possible using poly-SiC. In the case of an interleaving integration for creating a poly-Si MEMS insertion process, a high temperature annealing step is required. The annealing time increases with the thickness. The ability to use thicker microstructures is highly advantageous for electrostatic MEMS sensors as the capacitive coupling increases with the thickness of the microstructure. Therefore, in polysilicon MEMS, the electronics thermal budget limits the allowable MEMS device layer thickness.

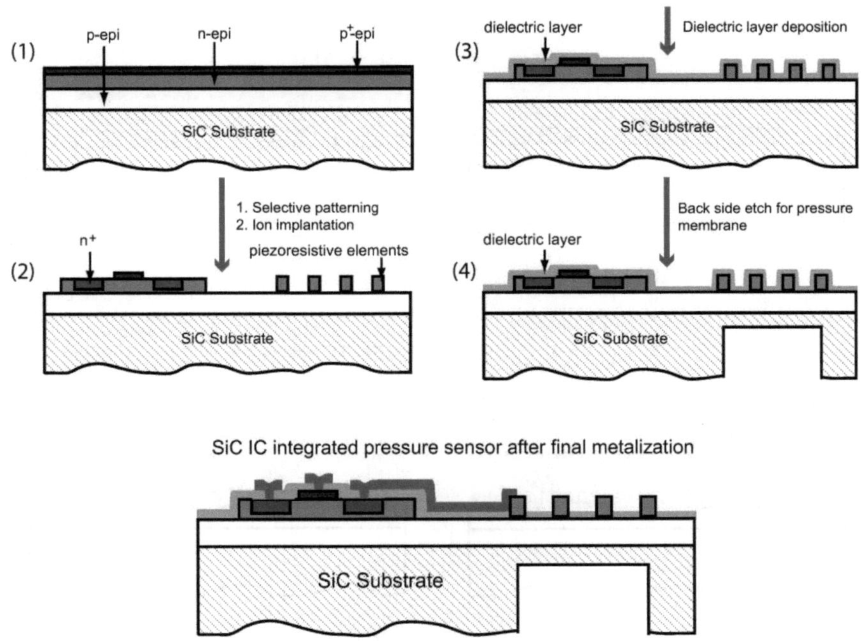

Fig. 6.3 Schematic representation of fabrication process sequence of SiC IC and piezoresistive pressure sensor integrated using an interleaved integration approach.

The interleaved approach can also be applied for creating MEMS, particularly MEMS membrane structures for pressure and acceleration sensing, using bulk micromachining. The fabrication process sequence used to create an integrated piezoresistive pressure sensor is shown schematically in Figure 6.3. Similar to the process described in Figure 6.2, the process starts with SiC substrates with epitaxial layers. The first step is to define electronic and piezoresistive elements by selective patterning. In this case, the n epi layer is not only part of the electronic structure but also part of the sensing element for diaphragm deflection. After patterning, the n^+ region is created by ion implantation. This step is followed by deposition of a dielectric layer, which acts as an insulating base layer for metal interconnects as well as a protective layer for electronics during bulk etching of SiC to create the membrane structure. The membrane structure can be created using DRIE or a combination of DRIE and electrochemical etching [19, 3]. Finally, metallization is performed to create electrical interconnects and MEMS-IC interface.

6.1.2.2 Post-IC MEMS (MEMS-last) Insertion for SiC

Post-IC insertion of SiC MEMS follows a very similar path of post-CMOS MEMS insertion used in Si technology. A simplified process sequence of SiC MEMS in-

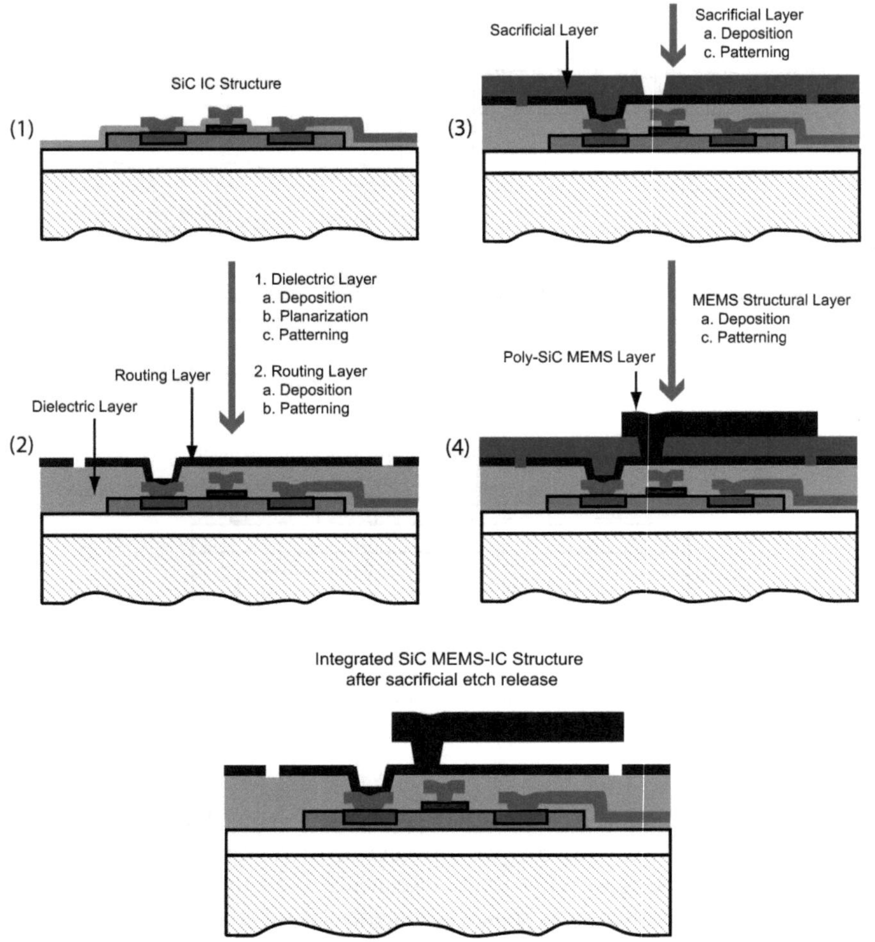

Fig. 6.4 Schematic representation of fabrication process sequence of SiC MEMS insertion of pre-fabricated SiC ICs.

sertion on SiC ICs is depicted in Figure 6.4. The process starts with the deposition and planarization of a dielectric layer on top of a pre-fabricated IC. This dielectric layer is patterned to create a via to the MEMS interface. This step is followed by the deposition and patterning of a routing layer. Afterwards, a typical SiC MEMS fabrication process can be followed, similar to one described in Figure 6.2. One advantage to this method is that IC and MEMS can be done in two dedicated foundries and the processes can be developed independently. Also, it may be possible to utilize the same layout space for both MEMS and the ICs, decreasing cost and system size.

Post-IC MEMS integration is perhaps the most straightforward way for SiC MEMS-IC integration. However, there are many process and reliability issues that have to be addressed before attempting this approach. One key factor is the metal

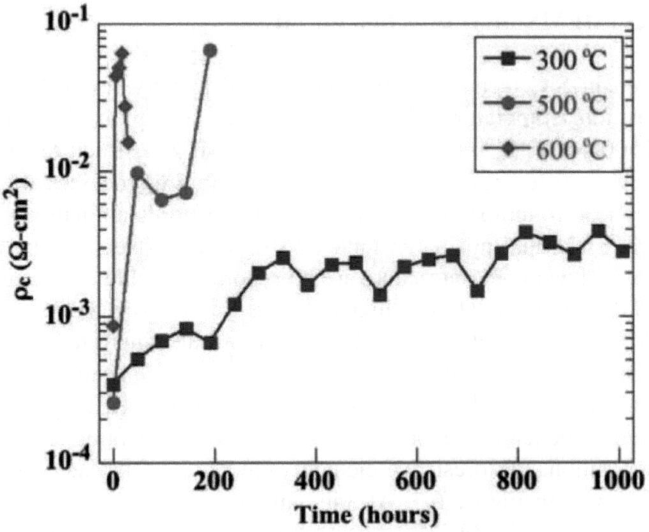

Fig. 6.5 [22] Average contact resistance of Pt/TaSi$_x$/Ni/SiC as a function of time at various temperatures determined by current-voltage measurements that were taken at room temperature, after cooling the samples (©Springer 2009), reprinted with permission.

contact degradation due to the high temperature exposure (800-900°C) during poly-SiC deposition on metalized ICs. There are many factors that affect the integrity of metal contacts in SiC devices, including metal type, temperature, and ambient environmental gases [20, 21, 22]. Most studies discuss the degradation of ohmic contact in air up to 600 °C. Figure 6.5 graphically represents the thermal stability of Pt/TaSi$_x$/Ni/4H-SiC ohmic contacts as a function of time at various temperatures. The data clearly demonstrates the pronounced aging of contacts due to very high temperature exposure [22]. As previously mentioned in Chapter 3, a report from NASA Glenn Research Center shows highly reliable metal contacts which use Pt/TaSi$_2$/Ti/SiC, and work thousands of hours at 500 °C without degradation. With regard to post-IC integration of MEMS, studies have to focus on the thermal stability of interconnects under deposition conditions for poly-SiC, namely the presence of precursor gases at temperatures up to 900°C. Another area of concern is the deposition of the routing layer, typically poly-SiC, on a metal surface. To the authors' knowledge, there are no reports on depositing CVD SiC on metal surfaces that have the sound electrical characteristics needed for the MEMS-IC interface. This should be one focus area of future research. For successful implementation of post-IC MEMS insertion for SiC, these interfacial issues have to be addressed from all fronts including materials, processes, and device design.

6.1.2.3 Pre-IC MEMS (MEMS-first) Insertion for SiC

Pre-IC MEMS insertion in SiC technology may be limited to very few MEMS structures as the deposition temperature of electronics grade SiC epitaxial layers are extremely high (1500-1600 °C, see Chapter 2). Typical MEMS structures involve many interlayers with different CTEs so that high temperature processes after MEMS fabrication may lead to temperature-induced cracking and delamination of some layers. Furthermore, thermal mismatch induced stress may adversely affect the quality of the epitaxial layers. To date, there are no reports available on the integrity of a SiC MEMS layer stack up at these high temperatures. However, one possible scenario of MEMS insertion before IC is fabrication of bulk-micromachined piezoresistive diaphragm pressure and acceleration sensors. Fabrication can start by defining the diaphragm using backside etching of the SiC layer by DRIE or combination of DRIE and electrochemical etching. The generic IC fabrication should begin with epilayer deposition. Piezoresistive elements have to be fabricated during the IC process as shown in Figure 6.3. An inherent advantage to fabricating diaphragm based sensors is that the MEMS process flow can be integrated to the IC fabrication sequence because only a one step etching process is involved; however, wafers have to be decontaminated prior to them being introduced into the IC fabrication process flow. Typical SiC DRIE uses metal masks, which can severely contaminate the IC tools and degrade IC peformance. Therefore, this approach may be impractical in most cases.

6.2 Power

Powers is obviously a vital aspect of a microsystem as operation of all the components — electronics, MEMS, and signal transmission — require power. Supplying power to a microsystem is rather complex since MEMS devices often have special power requirements. Some devices require high voltage while others demand high input power. In some instances, additional power connections and voltage-current conversion circuitry are also required. In many applications, supplying power to microsystems is not an issue when direct access to the outside world is feasible. However, most harsh environment applications such as combustion monitoring, structural heath monitoring in gas turbines and nuclear power plants, and space and military applications, may require on-board power and power management because access to the outside world is restricted. Furthermore, when arrays of distributed sensors are used, on-board power reduces the complexity of the electrical interconnects. Long term unattended operation, typically encountered in space applications, military surveillance, and structural health monitoring, requires on-board power generation, storage, and management. In general, power supply needs for microsystems are overlooked; most research is focused on MEMS and electronics. In the case of harsh environment applications, it may demand a completely new set of power generation or storage units that can withstand harsh ambient conditions such as high temperatures, high radiation, and corrosive media. Hence, hurdles to realizing successful remote power for harsh environment applications are significant.

The power consumption of microsystems mainly depends on the number of components present and on intended functionality. With regards to MEMS devices, most power may go to power conversion and signal processing electronics. If on-board wireless transmission is needed, perhaps it may be the most power hungry component of the system. At the system level, interconnect induced capacitance, contact and line resistance also increases power consumption. Therefore, an on-board power module should have the capability of handling all the power needs. Moreover, consideration should be given to the lifetime, size, weight, and environment compatibility of the power module based on the intended application.

Three main strategies are being applied or pursued for powering MEMS. Those can be categorized into standalone power sources, environmental power scavengers, and remote powering. Standalone power sources include batteries, micro-engines, micro fuel cells, and radiation-based power sources (betavoltaics). Typical environmental power scavengers are photovoltaic, thermoelectric power generators, and vibration energy harvesters. Remote powering generally applies inductive coupling or radio frequency transmission. Most of these power solutions are available at the macro scale. For many applications, the large form factor of these power devices may not negatively impact the feasibility of the application. However, some applications such as health monitoring of components of a gas turbine, sensors for downhole data logging, and space applications require smaller form factor systems. Thus, miniature power modules that are comparable in size to MEMS and electronic modules are highly desired. At the current stage of the technology, the power output of micro-scale batteries and transducers is extremely low. Research has to be focused on both power generation and power management in order to satisfy the anticipated microsystem needs. The problem has to be addressed from both a design and materials stand point to increase the output of these power supplying units.

6.2.1 Standalone Power Sources

Standalone power sources generate power through the conversion of stored chemical energy or radiation. One of the most widely used standalone power sources is the electrochemical battery. Other commonly considered standalone power sources are micro fuel cells, micro engines, and radiation-based power sources. Standalone power is useful for harsh environment applications in most cases. External power through wires is not always feasible. Each of these standalone power sources will be introduced, and considerations for implementation in harsh environments will be discussed.

6.2.1.1 Batteries

Electrochemical battery technology is a mature field and many battery types are commercially available. Portable electronic devices such as calculators, cell/smart phones, and portable computers are powered by electrochemical batteries. Depending on the electrochemical makeup of the battery, it may be rechargeable or non-rechargeable. High power density is highly attractive, and batteries compare well to other technologies. Furthermore, batteries provide direct current (DC), therefore eliminating the need for having complex power conditioning circuitry. When alternating current (AC) power sources are used, they typically require additional electrical components to convert AC to DC for microsystem applications [23].

In most cases, the size and the weight of the battery dominate the total weight and the volume of the system. Application scenarios demand low weight, small form factor microsystems. Many efforts have been focused on developing thin film battery technology to specifically address the needs of small form factor microsystems. The possibility of fabricating the battery directly onto or on the backside of the substrate with MEMS and IC is very attractive as it simplifies the interconnect complexity [24]. Furthermore, the operating temperature of thin film solid state batteries can be as high as 150 °C — much higher than the highest possible operating temperature of conventional lithium-ion batteries [25, 26]. A few companies have commercialized or are very close commercializing thin film battery technology; however, the predicted power density of these technologies are still not sufficient for producing microscale batteries that can serve microsystem technology [27].

To reduce the large planar area of thin film batteries and to increase the power, some research efforts have been shifted to stacking multiple cells to limit the footprint with a stack consisting of a small number of thin film batteries connected in parallel or serial manner. The projected energy for this kind of bundle is approximated to be 5 mWh/cm^2 at 3.5 mAh/cm^2. Additionally, research into novel chemical combinations to increase the achievable power density over conventional material sets is ongoing [23]. For instance, silicon-zirconium-silver has enhanced capacity retention during cycling while nickel metal hydride cells exhibits increased values of power and energy/unit area.

Despite their wide spread usage and success, there are a few drawbacks to battery power. Whether it is thin film or meso-scale batteries, they contain a finite amount of energy and have a limited shelf-life. Furthermore, all state of the art battery technologies have limitations with regard to operating temperature because the chemical components degrade. This limits their use as a power source for many harsh environment applications discussed in Chapter 1.

6.2.1.2 Fuel Cells

A fuel cell is an electrochemical cell that converts fuel into electrical energy through a chemical reaction. Hydrogen is a common fuel and methanol is also gaining momentum as an alternative. As shown in Figure 6.6, a fuel cell consists of a cathode

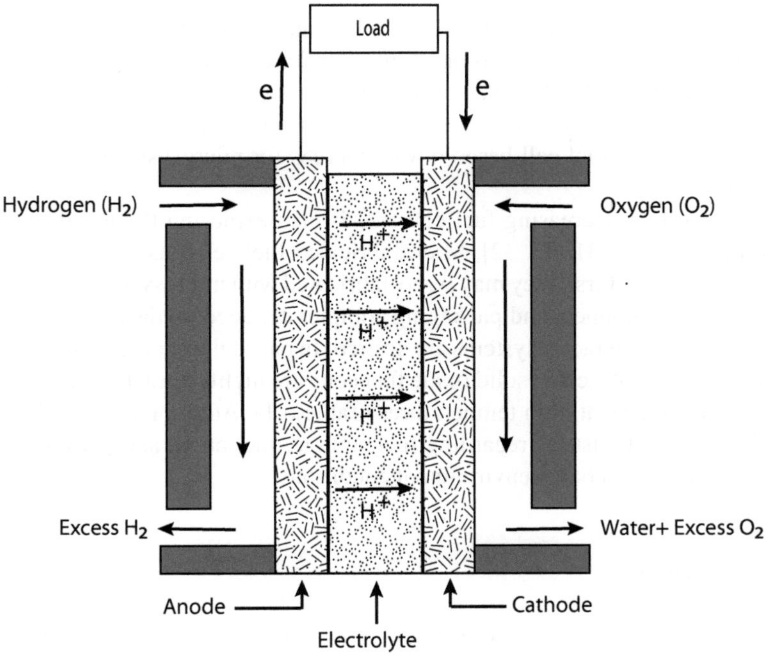

Fig. 6.6 Schematic representation of a hydrogen fuel cell.

Table 6.1 [28] A summary of common fuel cell types listed with their membrane, operating temperature and electrical efficiency.

Fuel Cell Type	Common Electrolyte	Operating Temp. (°C)	Electrical Eff.
Polymer Electrolyte Membrane	Solid organic polymer-polyperflurosulfonic acid	50-100	∼55%
Alkaline	Aqueous solution of KOH	90-100	∼60%
Phosphoric Acid	Liquid H_3PO_4	150-200	40%
Molten Carbonate	Liquid solution of lithium sodium and/or potassium carbonate	600-700	∼45%
Solid Oxide	Yttria stabilized zirconia	600-1000	∼40%

and anode separated by an electrolyte. Depending on the type of electrolyte, fuel cells can be categorized into several groups. These fuel cell categories are polymer electrolyte membrane (also known as a proton exchange membrane), alkaline, phosphoric acid, molten carbonate, and solid oxide. Depending on the cell type, the operational characteristics and power conversion efficiencies differ significantly. Table 6.1 summarizes the characteristics of common fuel cell types [28].

Fuel cell technology is very mature and has found applications in automobiles, space exploration, and as small-scale power plants. Miniaturizing fuel cells for portable device applications is a widely discussed topic. Many commercial industries are developing miniaturized fuel cells that are comparable in size and power density to conventional battery technology [29, 30]. These miniaturized fuel cells are potential replacements for lithium ion batteries in laptop computers and cell phones. Perhaps this market-driven technological push for achieving portable fuel cells will lead to suitable fuel cell based power sources for microsystems applications very soon.

Research into further decreasing fuel cell size is also exploring the use of microfabrication techniques [31, 33, 32]. Microfabricated fuel cells may have added benefits to microsystems. First, they may be co-fabricated with microsystem components reducing the interconnect and packaging complexity. Second, they are highly attractive when small form factor systems are desired. In particular for high temperature applications, microfabricated solid oxide fuel cells are highly attractive because they are capable operating at high temperatures (Table 6.1). Additional research is needed to address specific issues regarding fuel storage, fuel ports, and packaging that can likewise withstand harsh environment conditions.

6.2.1.3 Micro Engines

Miniaturized power generating devices using combustion have been considered for power needs in the range of milli-watts to watts. Research has focused on both millimeter scale and micro-scale devices [34]. Millimeter scale engines have shown to operate with good combustion efficiency and produce sufficient power for microsystem applications [35]. Micro-scale combustion power generators are based on microfabricated gas turbine or internal combustion rotary (Wankel-type) engine concepts [36, 34]. Although these micro-scale combustors are shown to produce positive power, many fundamental and engineering issues hinder their advancement. These issues include combustion in small volumes and surface to volume ratio impacting heat loss in the combustion chamber. Additionally, wear and corrosion of microfabricated engine components, efficient mixing of fuel with air inside micro-channels, and fabrication complexity are also challenges [34]. Although the practicality of micro scale combustors is limited, millimeter scale combustion engine concepts should be exploited. Rotary (Wankel-type) engine may open unique opportunities, specifically automotive and aerospace application mainly because of their fuel flexibility.

6.2.1.4 Radiation-based Power Sources

Radiation-based power sources typically use beta particles emitted by radio isotopes to produce electricity. The device use for this conversion is known as betavoltaic, which, in principle, is very similar to photovoltaics (see section 2.2.2.1). In contrast to photovoltaics, electron-hole pairs generation in betavoltaics occurs through the interaction of high-energy β-particles instead of photons. Radio isotopes, such as Ni-63 and tritium (H-3), have been considered as a beta source for low power batteries [37, 38]. The long half-life of these isotopes makes them ideal as long-life power sources. Furthermore, their insensitivity to environmental conditions is attractive for harsh environment applications [47]. Many semiconducting materials such as gallium phosphide [39, 40], aluminum gallium arsenide [41], amorphous silicon [42], porous silicon [43], and gallium nitride [44] have been investigated as beta collectors. Theoretical energy conversion efficiency of a betavoltaic material increases with increasing semiconductor band gap [45]. Thus, SiC has drawn considerable interest as a betavoltaic material due its wide band gap. Furthermore, the radiation hardness of SiC bodes well for reliable long-term operation [46].

Although very few studies have been reported on radio isotope SiC betavoltaics, the results are promising. Chandrashekhar *et al.* reports open circuit voltage of 0.72 V and a short circuit current density of 16.8 nA/cm^2 from a single p-n diode structure betavoltaic cell [47]. The diode was fabricated using p-type 4H-SiC with a 0.037 GBq Ni-63 source. A p-i-n diode betavoltaic built on a n-type 4H-SiC substrate with a 8.5GBq P-33 source achieves an open circuit voltage of 2.04 V with a short circuit current density of $2.1\mu A/cm^2$. As the voltage and current produced by these single cell betavoltaic devices may not be sufficient for many microsystems, a multi-device architecture has been proposed. One possibility of increasing power would be to use a stack of betavoltaic devices interleaved with high specific activity isotope layers, such as Ni-63, connected in a series or parallel fashion [48].

Considering the current state of all of these standalone power sources, batteries are the natural choice. Battery technology is very mature, widely available, and capable of providing sufficient energy for many microscale system needs. Miniature fuel cells may become a sound alternative in near future due to high power density and suitability for integration into microsystems. Likewise, transitioning miniature internal combustion engines into microsystem power sources requires extensive research. Betavoltaics currently suffer from low power output.

A common factor to all these standalone power sources is that they contain finite amount of energy. Batteries have to be replaced or recharged while both fuel cells and miniature combustion engines have to be refueled. In applications where power supplies cannot be recharged or replaced and long-term unattended operation is required, alternative approaches are needed.

6.2.2 Environmental Power Scavengers

Environmental power scavenging is a widely discussed topic for powering microsystems. Mainly, three different forms of environmental energy are considered: light, heat, and kinetic. With regards to the availability of environmental energy, harsh environment microsystems possess a unique advantage as many of the intended applications typically have one or more forms of the aforementioned energy sources. Temperature gradients and vibration are present in gas turbine and automotive engine environments as well as in many chemical processing scenarios. Solar energy and thermal gradients are common for space environments. Many different environmental power scavengers are used for converting these energy forms to electricity, including photovoltaics, thermoelectric power generators, and vibration energy harvesters. These energy scavengers can be an alternative to common power sources such as batteries or they can be used for extending the lifetime of common power sources. Thus, environmental power scavenging eliminates the need to replace batteries or at least greatly extend their lifetime, which facilitates long-term operation of the microsystem in harsh, often isolated, environments.

6.2.2.1 Solar Power

A solar cell, or photovoltaic, is employed to convert solar energy to electrical energy. A photovoltaic cell consists of a semiconductor p-n junction. When n side of the photovoltaic is subjected to solar light (photons), electron-hole pairs are generated within the depletion region where the built in electric field pushes electrons to the n side and holes to the p side creating an open-circuit voltage. If the cell is connected to a load, as schematically shown in Figure 6.7, a current will flow from one region to the other through the load. Since light is commonly available in many application environments, solar power would be a sound alternative for powering microsystems. Solar power generation is a highly developed technology and has produced power modules for many commercial and space applications. Miniature solar cells have been used for many decades for powering small consumer electronic devices such as calculators.

There exists a vast knowledge base on devices, materials, and fabrication of solar cells. Thus, extending this knowledge to produce power modules for microsystems is relatively straightforward. From the microsystem technology stand point, solar cell technology is very attractive because solar cell fabrication is highly compatible with existing microfabrication methodologies. Various integration schemes can be implemented as the material option for solar power generators includes single crystal silicon, polysilicon, amorphous silicon, and gallium arsenide. All are heavily used in microfabrication environments.

Many successful attempts on using miniature solar cells for MEMS and microsystems have been reported. Lee *et al.* fabricated a miniaturized high voltage solar array using amorphous silicon to power electrostatic MEMS [49]. In this work, an array of 100 single solar cells with total area of 1 cm^2 produced an open circuit voltage of 150 V and short circuit current of 2.8 μA. This array has been used to move an electrostatically-driven Si mirror. A big step towards achieving on-board power for microsystems occurred in 2003 when a fully-integrated microsystem fabricated using SOI was demonstrated [50]. The system was composed of an electrostatic gap-closing actuator powered by an on-board solar cell array with the power being switched by an on-board buffer consisting of an NMOS inverter. The solar cell array consisted of 200 cells, each with dimension of 400μm by 400μm, and were able to produced an open circuit voltage of 85 V.

Although power supplied using solar cells may provide the necessary power for many microsystems, solar cells are not well suited for environments that do not have access to light. Furthermore, the current silicon technology is also not suitable for applications requiring exposure to high temperature or high levels of radiation. NASA has started exploring wide bandgap solar cells, including cells from GaInP, GaP, GaN, and SiC, specifically to address harsh environment applications [51]. These materials have exhibits both radiation hardness and high temperature stability; however, other factors affecting the lifetime of the device such ohmic contact and interconnect degradation still need to be solved.

6.2.2.2 Thermoelectric Power

When a thermoelectric material is subjected to a temperature gradient, heat flows from the hot end to the cold end and releases carriers (electrons or holes) that likewise diffuse from the hot end to the cold end. This phenomenon, known as the Seebeck Effect, is used for temperature measurements as well as power generation. Thermoelectric power generation is typically achieved by connecting heavily doped n-type and p-type semiconductors connected electrically in series but connected thermally in parallel. This device structure is commonly referred as a thermocouple. Temperature gradient forces dominant charge carriers of each material, electrons in n-type and holes in p-type, to diffuse towards the low temperature side of the each leg; producing a voltage different across the two sides. The power output of a thermocouple depends on thermoelectric material and temperature gradient. For typical applications, several thermocouples are connected serially to increase the output power. Most commercially available thermoelectric generators, commonly known as thermopiles, contain as many as 128 thermocouple units. A generic configuration of a thermoelectric power generation system is schematically shown in Figure 6.7.

The efficiency of thermoelectric materials can be expressed as a dimensionless parameter, ZT, given by Equation 6.1:

$$ZT = \alpha^2 \bar{T} / \rho \lambda \tag{6.1}$$

Where α is the Seebeck coefficient (ΔV/ΔT), ρ is the electrical resistivity, λ is the thermal conductivity and $\bar{T}$ is the average absolute temperature of the hot and cold interface temperatures.

High efficiency is achieved by maximizing ZT. Consequently, effective thermoelectric materials should have a high Seebeck coefficient, low thermal conductivity, and low resistivity. Bismuth Telluride (Bi_2Te_3) alloys and PbTe alloys are two of the most common thermoelectric materials used today. Bi_2Te_3 has a ZT value of around one at room temperature. On the other hand, PbTe reaches ZT of one at 500-700K [52] suggesting that PbTe is well suited for relatively high temperature applications. SiGe is one of the highest temperature thermoelectric materials as it can operate up to 1000 °C [53]. The major drawback to SiGe is its low ZT (<1) throughout the entire temperature range. Efforts have focused on engineering materials with higher ZT for high efficient thermal power generation. One successful approach reported by Venkatasubramanian *et al.* shows ZT of 2.4 in thin film p-type Bi_2Te_3/Sb_2Te_3

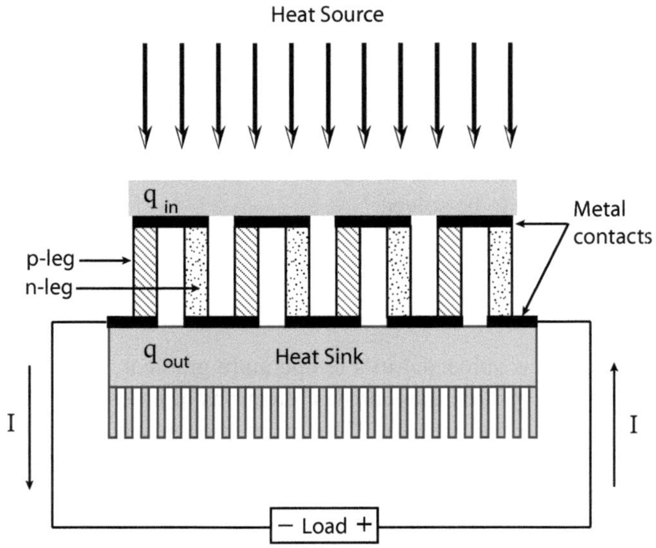

Fig. 6.7 Schematic drawing of a thermoelectric power generation system.

Table 6.2 ZT of poly-Si, SiGe and SiC with corresponding temperature

Thermoelectric Material	Figure-of-merit (ZT)
p-doped Polysilicon	0.005 at 300K
n-doped Polysilicon	0.012 at 300K
p-doped PolySiGe	0.037 at 300K
n-doped Poly SiGe	0.061 at 300K
Nitrogen-doped SiC films (sintered)	0.125 at 973K
Boron-doped SiC films (sintered)	0.021 at 973K

semiconductors. This material system appear to have a unusual structure, a superlattice formed by alternating layers of Bi_2Te_3 and Sb_2Te_3 semiconductors, which results in low thermal conductivity [54].

In addition to conventional III-V materials, CMOS compatible materials such as poly-Si, SiGe, and SiC have also drawn considerable attention as thermoelectric materials for micropower generators due to their compatibility with MEMS and IC fabrication. For instance, a thermopile made from an array of microfabricated n-polySi/p-polySi thermocouples in a 1 cm by 1 cm area achieves an open-circuit voltage of 16.7 V and output power of 1.3 μW when a 5K temperature gradient is maintained. An investigation of a microfabricated n-poly-SiC/p-poly-Si thermopile exhibits higher output voltage than that of n-poly-Si/p-poly-Si and Al/p-poly-Si thermopiles [55]. In terms of process integration, these materials holds advantage over tradition thermoelectric materials; however, the conversion efficiency of these materials are quite low (Table 6.2) [56, 57].

From an industrial prospective, micropower generation using thermoelectric materials has grown considerably [75, 59]. There are many micro-thermoelectric generators now available on the market, and the power output of these devices varies widely depending on the number of thermocouples present, device geometry, and temperature gradient. Thus, thermoelectric power generation units, despite their relative size, can play a significant role in microsystem technology, especially when a sizable thermal gradient is present. For example, power for structural health monitoring of steam tubes in power plants are a good fit for thermoelectric power generation since the wall temperature is typically higher than that of ambient, setting up a natural gradient.

6.2.2.3 Vibration Power Scavenging

Vibration energy harvesting is a highly active area of research, and there is ever-growing interest in using it for wireless, self-contained electronics as well as microsystems. Many harsh environment applications are subject to sustained vibrations, so assessing the potential of harnessing this energy for powering microsensors is of particular interest. For instance, a car engine can vibrate up to 200Hz with 12 ms^{-2} acceleration [23]. There are three major transduction mechanisms for vibration power scavenging: piezoelectric, electromagnetic, and electrostatic. Each of these techniques has its own advantages and disadvantages in terms of power density, device architecture, operational requirements, and ability to integrate with the other components of a microsystem [60].

Piezoelectric materials become electrically polarized when they are subjected to mechanical strain. The degree of polarization is proportional to the applied strain. The relationship between polarization and the applied strain is specific to the piezoelectric material and its crystal structure, and the magnitude of the response is given by the piezoelectric coefficients of the material. Piezoelectric materials typically exhibit anisotropic characteristics, consequently, the degree of polarization is also depends on direction of applied strain with respect to the orientation of the polarization. Anisotropic properties of piezoelectric materials are defined by a series of symbols and notations. Based on these definitions, the compressive strain applied perpendicular to the electrodes utilizes the d33 coefficient of the material while transverse shear strain applied parallel to the electrodes employs the d31 coefficient for power generation (Figure 6.8). For most vibration power scavenging schemes, transverse shear strain is utilized because compressive coupling is not feasible [61].

The most common device configuration exploited in energy harvesting using the d31 mode is the cantilever beam. This structure is suitable for many energy harvesting applications because it provides a relatively large strain under small input force. Furthermore, cantilevers typically operate at low resonant frequencies and can be easily tuned to desired frequency by changing the geometry. Both characteristics are highly advantageous as typical vibration frequencies in real world systems range from one to a few hundred Hertz [23]. The resonant frequency can be further reduced by attaching a proof mass at the end of the beam. This will also increase the

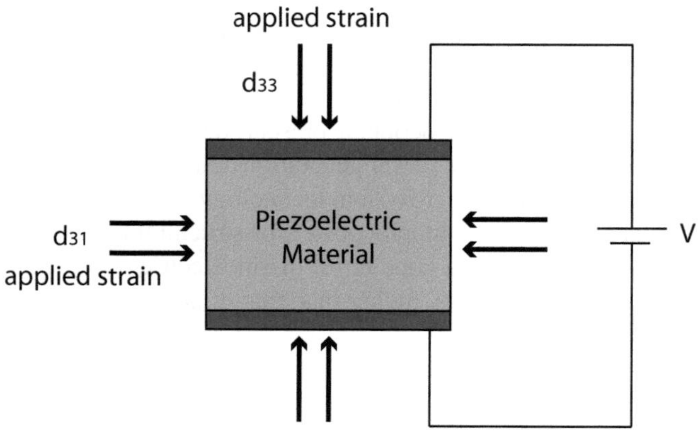

Fig. 6.8 Piezoelectric power generation through applying compressive transverse shear strain.

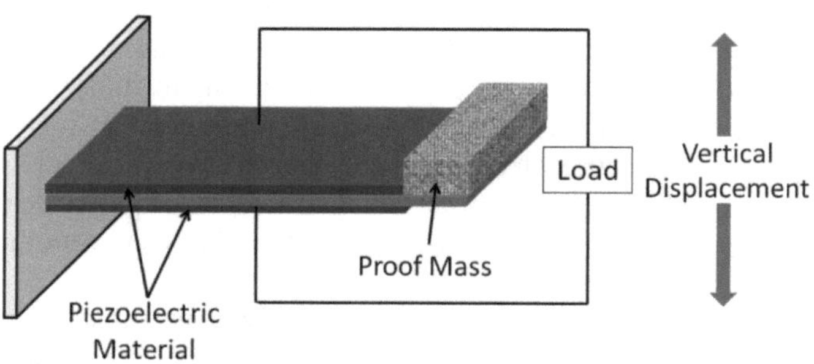

Fig. 6.9 [62] Schematic drawing of cantilever-based piezoelectric power generator which uses the d31 mode (©Springer 2009), reprinted with permission.

deflection, resulting in higher strain, thereby increasing the output power. A typical cantilever beam structure used in piezoelectric energy harvesting is schematically shown in Figure 6.9 [62]. Piezoelectric elements are either deposited or bonded to the substrate.

The power output of these cantilever energy harvesters differs with the materials, electrode geometries, and structural resonant frequency. There are many materials, from single crystalline to polymeric, that exhibit piezoelectric behavior. Some of the notable ones include quartz, lithium niobate ($LiNbO_3$), lead zirconate, lead titanate ($Pb[Zr_xTi_{1-x}]O_3$), and polyvinylidene difluoride (PVDF). However, a majority of piezoelectric power scavenging devices are based on lead titanate (PZT) due to its mature fabrication methodology. Furthermore, it has a relatively larger piezoelectric coefficient which in turn provide more power output for a given input acceleration [63]. Aluminum Nitride (AlN) is also gaining wide attention from the MEMS community because its fabrication methodologies are compatible with conventional microfabrication technology. However many AlN-based piezoelectric devices to date are mainly limited to resonators for frequency referencing and filtering [64, 65, 66]. No report in the literature are available on AlN based vibration power scavenging devices. The reason may be attributed to the small piezoelectric coefficient.

Most piezoelectric materials lose their piezoelectric property when the temperature exceeds a specific point, commonly referred as the Curie temperature (T_c). Many piezoelectric materials has relatively low Curie temperatures [67]. For instance, the T_c of PZT is around 350 °C, leading to a safe operating range between 150-200 °C. Thus, power generation in a high temperature environment requires materials with high T_c. There are many commercially available piezoelectric materials that possess high Curie temperatures. These include bismuth titanate (T_c = 600 °C) and lithium niobate (T_c = 1210 °C). However, the piezoelectric response of these materials is relatively low. With proper packaging technology, existing piezoelectric materials including PZT may suffice for many harsh environment applications, excluding high temperature. Further research is required in high temperature piezoelectric materials as well as fabrication methodologies in order to push vibration power scavenging for demanding environment applications.

Vibration power scavenging using microscale piezoelectric devices is still at the early research stage, though their macroscale counterparts are commercially available. Some commercially produced devices are capable of producing power ranging from a few microwatts to few mlliwatts [68]. One such a commercial device is Volture^{TM} PEH 20W. This device is 92 mm by 44 mm by 10 mm and produces up to 9 mW at a vibration frequency of 150 Hz. For autonomous microsystems, these commercial power scavenging devices are attractive; however, the harsh environment compatibility of these devices may restrict the application space. The larger size of these commercial devices may also be a limiting factor for some applications.

Vibration energy harvesting using electromagnetic transduction is also an active area of research for MEMS and electronics power needs [60]. An electric current is generated when a magnetic flux oscillates through an inductor. This phenomenon was first discovered by Faraday and is one of the most widely used techniques for large scale power generation. Two device configurations are used for vibration energy harvesting using an electromagnetic effect: a fixed magnet whose flux is coupled with a moving induction coil or a moving magnet whose flux is linked to a fixed coil. The latter is typically used for vibration power scavenging as it simplifies the electrical connections to the coil. Similar to piezoelectric energy harvesters, resonating cantilever beams are widely used in electromagnetic power scavengers. Figure 6.10 schematically represents the generic device architecture.

The power generated by these devices is proportional to the magnetic flux density and magnitude as well as number of turns in the induction coil. With regards to macroscale devices, high performance magnets and multi-turn coils are readily available. Commercially available electromagnetic vibration energy harvesters have shown to produces up to power of 147 mW at 11 V output voltage [70]. One such device generates about 3 mW AC power at 0.5 m/s^2 RMS acceleration (Figure 6.11).

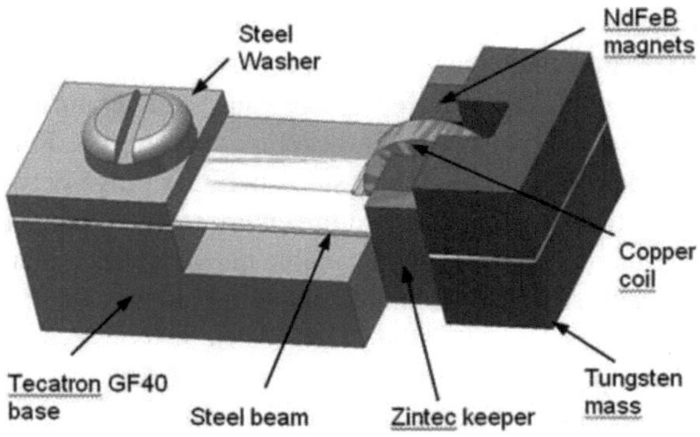

Fig. 6.10 [69] Schematic drawing of cantilever-based electromagnetic power generator (©Springer 2007), reprinted with permission.

Fig. 6.11 [69] Perpetuum PMG7 electromagnetic vibration power harvester (©Springer 2007), reprinted with permission.

This type of device has found use in various industrial settings such as on railcars for powering various types of sensors and wireless modules.

MEMS scale electromagnetic power generators has also been fabricated in order to reduce the form factor of the device. Microscale magnets and coils can be fab-

Table 6.3 Properties of commercially available magnets [62]

Magnetic material	Curie temp. (°C)	Maximum working temp. (°C)	Flux density (mT)
NiFeB (N38H)	320	120	450
Ceramic	460	250	100
SmCo	750	300	350
Alnico	860	550	130

ricated using microfabrication techniques. However, microfabricated magnets have poor flux densities, and the number of turns that can be achieved with planar coils are limited. Thus, the power generated from microscale energy harvesters is very low, typically below 1 μW, insufficient for powering a typical microsystem. Furthermore, the amplitudes of vibration at the MEMS scale is also limited. Therefore, achievable power through flux variation is also limited [69].

NdFeB magnets are the preferred type of magnets for power generation due their high flux density (450mT); however, NdFeB magnets are not suitable in power generation at high temperature condition as this material has a relatively low Curie temperature, *i.e.*, 320 °C. As shown in Table 6.3, some commercial magnets have high Curie temperature, but their magnetic flux densities are relatively low ([62]). Vibration power scavenging electromagnetic generators using centimeter scale systems have been demonstrated. However, a considerable amount of work is required on both MEMS scale and high temperature compatible electromagnetic power generation.

Electrostatic transduction is another way of generating power from environmental vibrations. Typical electrostatic energy harvesting devices are comprised of two conductive plates that are electrically isolated via air, vacuum, or a dielectric insulator. These plates are charged by using an external source such as a battery. The capacitance (C) between these plates can be expressed as:

$$C = Q_E/V_E \tag{6.2}$$

where, Q_E is the charge on the plate in Coulombs and V_E is the voltage on the plates in volts.

Devices are designed such a way that these two plates can be moved relative to one another through vibration excitation. The electrostatic energy conversion can be accomplished using a fixed charge method or fixed voltage method. Both cases are divided into two phases. The first phase of the fixed charge method occurs while the plates are moving towards each other and the plates are isolated from the energy reservoir. At the start of the second phase, when the plates are closest together, the capacitors are connected to the reservoir in order to charge the plates but then again disconnected, fixing the charge in the system. During the second phase, the plates start to separate, forcing the voltage to increase, increasing the energy stored in the capacitor. This energy is a conversion of the mechanical energy needed to separate plates into electrical energy. Repeating the first phase, the plates again begin to come together. During this period, the capacitor is attached to a second discharge reservoir and allowed to freely discharge. Another set of circuitry is used to move the energy

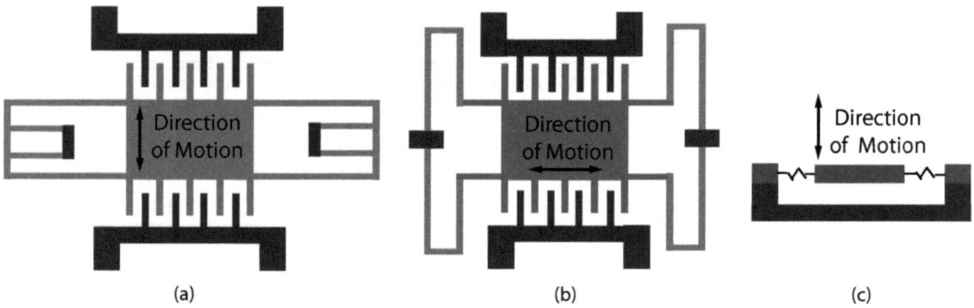

Fig. 6.12 [73] A schematic illustration of typical device architecture (a) in-plane overlap-varying, (b) in-plane gap-closing, and (c) out-of-plane gap-closing (©Elsevier 2003), reprinted with permission.

dumped into the discharge reservoir into the main energy reservoir, resulting in a net increase of energy stored in the energy reservoir during each cycle [71]. The first phase of the fixed voltage method operates in the same manner. However, during the second phase, while the capacitor plates separate, a fixed, high voltage is forced onto the capacitor that also serves to charge it initially. In order for the high voltage to be maintained as the plates separate and capacitance decreases, charge is dumped from the capacitor to the discharge reservoir. As the first phase repeats and the plates again come together, the fixed voltage condition is removed and the capacitor is allowed to discharge, returning additional charge to the reservoir. Because of added hardware complexity of implementing a fixed, high voltage configuration as well as additional loss mechanisms present in the fixed voltage method, a fixed charge configuration is more common [72].

Microfabricated electrostatic energy harvesting devices fall into three categories: in-plane overlap-varying, in-plane gap-closing, and out-of-plane gap-closing [73]. Schematic illustrations of typical device architecture for these three topologies are shown in Figure 6.12.

The use of electrostatic vibration power scavenging for microsystems is still in the research phase. The lack of development to date can be attributed to some inherent disadvantages of this method. One key drawback to electrostatic energy harvesting is the need for an input charge or voltage supply unit, which may not be practical in many applications. Furthermore, electrostatic power scavengers produce high voltage and low current. This may not be suitable for many device applications. The main advantage of using electrostatic devices for energy harvesting is the ability to integrate these devices into traditional microsystem fabrication processes; the energy scavenging structure can be fabricated using the same material as the MEMS device. Thus, electrostatic vibration power scavenging would allow the power generator to be fabricated in SiC as well as minimizing interconnects since a separate system component would not need to be integrated as part of the first or second level packaging. This makes it the best choice for vibration power scavenging for very high temperature and high corrosion applications, where a high level of monolithic integration is preferable for increased system reliability.

Table 6.4 Key features of converting vibration energy to power [60, 74]

Harvesting Method	Piezoelectric	Electromagnetic	Electrostatic*
Electrical output characteristics	High Voltage Low Current	Low Voltage High Current	High Voltage Low Current
Power density	25 mJ/cm^3	35 mJ/cm^3	4 mJ/cm^3
Feasibility of integrated fabrication	low	low	high

* Needs external charge source

All the vibration power scavenging methods discussed — piezoelectric, electromagnetic, and electrostatic — should continue to be researched as harsh environment power sources. Table 6.4 summarizes the key features of converting vibration energy to power. Each method has its own advantages and disadvantages. For instance, electromagnetic generation produces comparatively high output current levels at the expense of low voltages while electrostatic and piezoelectric energy harvesters generate high voltage and low current [60]. The power needs of the microsystem, operating environment, and integration capability are all critical when a selecting power harvesting method. A combination of one or more methods may be an option. This would also potentially counteract shortcomings of a given method.

6.2.3 Remote Powering (RF powering)

Remote powering would be another alternative for supplying power to microsystems when connections to the outside world is restricted, and the desired operational lifespan exceeds that of conventional batteries. Powering embedded microsystems remotely has been investigated for many medical implants as the replacement of a battery in implanted device typically requires surgery [75]. Remote powering is useful for other harsh environments as well. For example, using embedded sensors for structural health monitoring of gas turbine blades and corrosion monitoring in steam tubes are potential fits for remote powering.

Remote power transfer is most commonly accomplished using inductive coupling between a transmitting coil and a receiving coil integrated with the microsystem. The power transmitted through inductive coupling can be of the order of few ten of mW [76]. The power is sufficient enough to drive sensor circuits or microactuators as well as charging of onboard rechargeable batteries. Suster *et al.* reports a radio frequency (RF) powering system that produces a stable DC voltage of 2.8 V with a 2 mA current supply capability from a 50 MHz RF power source [77]. This system has been successfully used for powering a MEMS strain-sensing system. Commercial interest in remote powering of microsensors is also gaining momentum with some of these devices already in the market [75]. These examples demonstrate the remote powering of standard Si microsystems and should be adaptable to powering harsh environment microsystems as well.

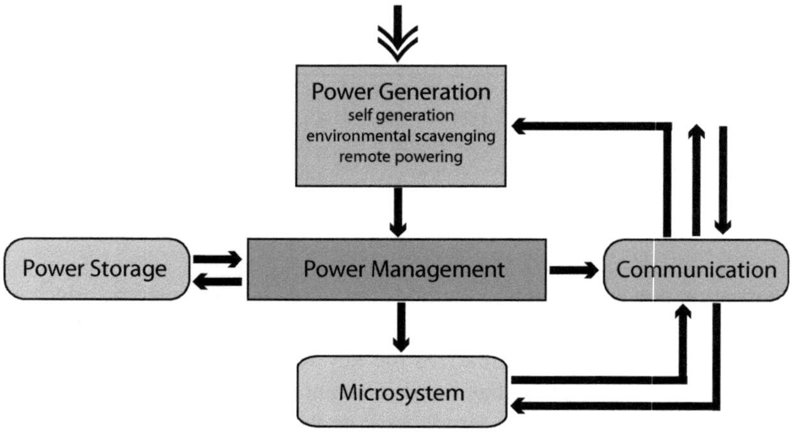

Fig. 6.13 A concept of self-sufficient microsystem with on board power generation, power management, and storage units.

Environmental power scavengers and remote powering can provide much needed on board power for microsystems; however, these power approaches suffer from low power density, fluctuation in energy level, and possible interruption of power. Therefore, onboard power management and power storage are also needed to capture the power when available and ensure uninterrupted, constant power to the system [78]. The power management system serves two different purposes. It converts AC or DC inputs from power generators to the desired form of energy for storage or operation. Furthermore, it also regulates the power distribution among various system components. A conceptual drawing of a microsystem that uses one or more regenerative power sources with power management and storage units is shown in Figure 6.13. Storage is typically accomplished using rechargeable batteries or capacitors, which, together with the power management circuitry, provides a smooth power supply to the microsensor even though the power generation occurs in a step-wise, nonlinear, or erratic fashion.

It is common that portable electronics and microsystems today rely on batteries. Limited lifetime restricts the application of batteries where replacement or recharging is not possible. High temperature compatibility and size scaling are current critical issues. The possibility of scavenging ambient energy through various energy harvesting technologies may remove some constraints on powering microsystems. Alternately, ambient energy harvesting enables extending the lifespan of conventional batteries. To date, these power scavenging technologies at miniature scale (cm, mm, or micrometer) are relative new; there are many challenges to be overcome. Challenges include increasing power output, development of high temperature compatible materials for energy harvesting needs, and creating viable manufacturing and integration schemes. Some of the challenges in remote powering include environmental disturbances and limitation in the distance between the source and the receiver. For sustainable harsh environment compatible microsystems, one of the key areas required immediate attention is power.

6.3 Communication

Similarly to power requirements, communication to the microsystem using wires is not always feasible, especially for harsh environment applications. For instance, if the environment consists of harsh chemicals, a wire electric contact could be the weakest point of whole system and may pose serious reliability issues. Furthermore, communication through wires is impractical when the microsystem is attached to a rotating body, such as turbine blades. In these situations, wireless communication is the only practical choice.

Many reports are available on RF communication to microsensors. For instance, RF powering and data transmission has been demonstrated for a microscale strain sensor [77], miniaturized drug delivery system [79], and harsh environment temperature sensor [80]. The strain sensor operating on a rotating shaft with RF link shows sensing resolution comparable to through-wire data transmission (0.09 $\mu\varepsilon$ over a 10 kHz bandwidth with 81 dB dynamic range) [77]. Currently, microscale strain sensors with RF signal transmission are commercially available. One such an example is the EmbedSense wireless strain sensor from MicroStrain, Inc. [75]. These examples show the feasibility and applicability of wireless communication to microsystems. However, to date implementing and testing these techniques using high temperature compatible materials such as SiC has yet to be demonstrated.

6.4 Testing of Harsh Environment Electronics, MEMS and Microsystems

Testing of harsh environment electronics, MEMS, and ultimately the microsystems that develop from integrating these components is one of the critical areas that can greatly impact the progress of SiC microtechnology. Replicating the demanding environmental conditions during testing requires extreme test strategies and suitable equipment. Many standard test setups for electronics and MEMS are limited to ambient and moderate environmental conditions. As such, they are not adequate for simulating some of the harsh environment applications outlined in Chapter 1. Therefore, exploring new test methods and suitable test fixtures are required. In addition to the lack of suitable equipment, limited support components that can withstand harsh environment conditions also complicates the testing issue. For example, testing of SiC MEMS at high temperatures needs high temperature compatible supporting electronics. SiC electronics are not at the stage to provide necessary functionality for MEMS operation, hindering high temperature MEMS testing. Due to these difficulties, most harsh environment tests to date are either passive or performed using custom built or retrofit equipment. Some of the notable harsh environment tests reported to date are briefly discussed below.

6.4.1 High Temperature Testing of SiC Electronics

Probe stations are commonly used to test discrete electronics at ambient conditions. Probe stations can also support short-term high temperature electrical measurements. However, long-term high temperature operation is more challenging. For instance, long-term high temperature operation leads to rapid degradation of physical and electrical properties of probes mainly due to oxidation of the probe tips and the contacts. That will introduce probe induced parasitic to the measurand, causing errors. Modifying standard test equipment and accessories to withstand high temperatures tends to be relegated to solutions that require in-house development. For example, NASA Glenn researchers have demonstrate high temperature (500 °C) testing of SiC junction gate field-effect transistor devices up to 7000 hours, to date the longest high temperature testing reported for SiC based devices [81]. SiC electronic chips were housed in a custom developed package using high temperature ceramic material as the substrate. The ceramic contains 96-wt% aluminum nitride and 90-wt% aluminum oxides [82]. The metal contacts were made using thick gold films. Then, the chip package was mounted on a high temperature compatible ceramic circuit board, and glass insulated Au wires were used to make connections to the testing apparatus. The testing was done in a bench top oven in ambient air. Figure 6.14 shows the prototype of packaging and ceramic circuit board developed at the NASA Glenn research center [83]. Though it is feasible to implement custom test strategies, development of standard test setups for high temperature testing would relieve a burden on researchers and accelerate the development of high temperature electronics.

6.4.2 High Temperature Testing for SiC MEMS

MEMS testing at high temperature is complicated by the fact that MEMS needs supporting electronics, which should withstand the same conditions. Currently, SiC MEMS device testing primarily must rely on expensive rack-mounted test equipment or build the required test board using Si electronics because SiC electronics are not readily available for certain applications. In the case of high temperature testing, rack-mounted test equipment is of limited usefulness. The use of silicon electronics is also challenging since MEMS devices need to be in very close proximity to the drive and sense circuitry to minimize interconnect induced parasitic capacitance. This adds complexity to the MEMS testing apparatus as the MEMS die has to be heated while keeping the Si electronics at suitable operational temperatures. Therefore, many active MEMS tests were limited to temperatures below 300 °C with a few exceptions, which have gone up to 600 °C using customized test setups. Even in these cases, the duration of testing was limited to only to several minutes.

One such a system, reported by researchers at University of California, Berkeley, uses an IR spot heater to locally heat the MEMS die while isolating the electronic

Fig. 6.14 [83] Four NASA SiC test chips in custom packaging (without any lids, exposed to air during all testing) and mounted on a custom high-temperature circuit board (©Wiley 2009), reprinted with permission.

circuitry [84]. Figure 6.15(a) shows the schematic cross-section of the test apparatus while Figure 6.15(b) shows a photograph of the printed circuit board (PCB) with wire bonded MEMS die. As shown in the figure, a hole is cut in an insulative foam and the MEMS die is positioned over the hole so that heating can be applied from the back side of the die. Then the foam is attached to the PCB such that the die is positioned into a cutout in the PCB ensuring no direct contact between the heated die and the PCB. A heat sink is used to further reduce heating of the adjacent components on the PCB. This setup enables heating of the MEMS die while keeping all of the PCB-mounted electronics at a cooler operational temperature. This set up has shown to work up to 600 °C. However, longer heating time eventually heats electronic components in the immediate vicinity due to heat transfer through the PCB. Therefore, testing is limited to one to two minutes. Though the experiment can be repeated many times, long term reliability testing at a sustained high temperature is not possible. This example clearly demonstrates the difficulties faced by high temperature testing of SiC MEMS and re-emphasizes the need for high temperature electronics.

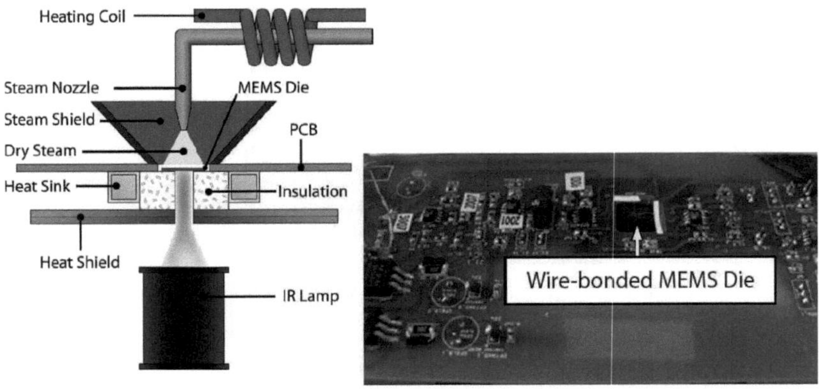

Fig. 6.15 [84] (a) the schematic cross-section of the test apparatus and (b) a photograph of the printed PCB with wire bonded MEMS die ((©SPIE 2009), reprinted with permission.

6.4.3 Shock Survivability Testing

Survivability and operation under high shock is important for many military and aerospace applications. In extreme cases, acceleration can exceed 100,000 g (g = gravitational acceleration, 9.8 m/s^2). Two methods have been used for applying a high-g shock to MEMS devices. One method uses a drop test, in which a drop table is accelerated towards an anvil using an elastic cord. MEMS substrates are attached to the drop table and the shock event occurs at impact with the anvil. This is a shot-pulse shock method and is generally limited to 40,000 g. Brown *et al.* have reported active testing of many MEMS devices under high g using a shock table [85]. Even though this method is limited to moderate shock events, one advantage to the shock table method is the availability of space for supporting electronics, allowing active testing of MEMS. The other shock testing method uses an air gun to create the shock event. In this case, MEMS die attached to a carrier body is launched using a high pressure air gun (Figure 6.16). This is a long-pulse shock method and the g-shock can exceeds 100,000 g [84]. For the air gun configuration, compact packaging of the entire system — MEMS, signal conditioning electronics, and power module — is needed if active testing is desired.

The first reported shock testing on a SiC MEMS device was performed on a SiC piezoresistive accelerometer [87]. Both the survivability and the operational characteristics were tested up to 40,000 g. Azevedo *et al.* report the survivability test on SiC DETF resonator using the drop test table [86]. Later, resonator survivability was verified up to 64,000 g using an air gun system [84]. Both tests were passive: devices were not in operation during the shock event. However, the functional tests following the shock event shows no changes to the performance of the MEMS device.

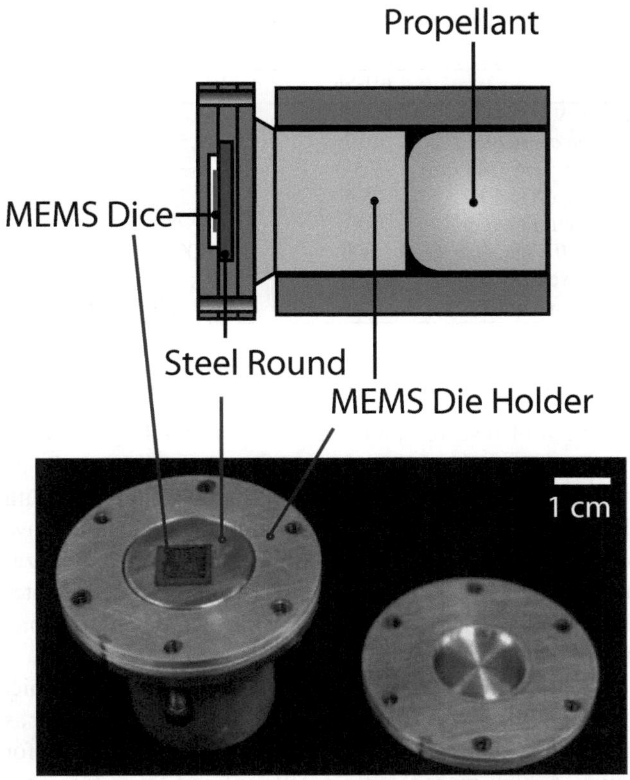

Fig. 6.16 [84] Accelerated drop test rig used for high g-shock survivability assessment (©SPIE 2009), reprinted with permission.

In some instances, real-world applications exhibiting significant shock or vibration occurs simultaneously in high temperature environments. Even though shock testing at ambient conditions has been accomplished using the previously discussed methods, it is extremely difficult to test in combined harsh environment conditions, such as heating and shock together, in a control manner. Thus, simulating some real harsh environment applications are difficult without first investing in testing technology. For example, creating a drop test that can be performed in a environmental control chamber.

6.5 Gallium Nitride, A Prospective Material for Harsh Environment Microsystems

Recent advances in processing and device technology of silicon carbide (SiC) have made promising inroads to practical harsh environment microsystems. This has in

Table 6.5 Comparison of key electronics properties of 2H-GaN to 4H-SiC polytypes at 300 K [88, 67].

Property	4H-SiC	2H-GaN
Bandgap (eV)	3.2	3.4
Relative dielectric constant	9.7	9.5
Breakdown field $N_D = 10^{17}$ cm^3(MV-cm)	2.5-3	2-3
Thermal Conductivity(W/cm-K)	3-5	1.3
Intrinsic carrier concentration (cm^3)	10^{-7}	10^{-10}
Electron mobility at $N_D = 10^{16}$ cm^3 (cm^2/V-s)	800	900
Saturated electron drift velocity (cm/s)	2 x 10^7	2.5 x 10^7
Piezo-electric Coefficients (pmV^{-1})	N/A	d31= -1.3-1.9 d33= 2.3-3.9

turn opened up a new paradigm of research aimed at further expanding harsh environment compatible microsystem technology. Among several possible candidate material platforms, gallium nitride (GaN) have drawn considerable attention. With the GaN optoelectronic device market thriving, material quality and availability have improved significantly with regard to substrate size and thin epitaxial films. Realization of GaN based electronics is now possible. Extending this knowledge of material and fabrication to MEMS and microsystems deserves consideration as this may open a new research venue for harsh environment microsystems.

GaN is a direct bandgap semiconductor with remarkable piezoelectric, electronic, and optical properties. GaN exhibits a wide bandgap similar to SiC and is used as a substrate for high power and high frequency electronics and is being explored for high temperature electronics. A high piezoelectric coefficient and favorable optoelectronic properties make it an ideal candidate for both sensors and actuators. It is also a potential candidate for piezoelectric transducers for vibration power scavenging. Furthermore, GaN is chemically inert to many corrosive chemicals and, therefore, is attractive for MEMS operating in corrosive conditions [67, 90]. All these properties imply that GaN is a strong competitor to SiC as a material platform for harsh environment microsystems because, like SiC, all the system components can potentially be fabricated from GaN.

Table 6.5 compares key electronic properties of 2H-GaN (hexagonal) with 4H-SiC. As an electronic material, it is similar to SiC. Its wide bandgap along with low intrinsic carrier concentration, makes it well suited for high-temperature and radiation-hard electronics. GaN-based electronic devices have been extensively studied for high-power and high-frequency applications due to its wide bandgap, high-breakdown field, and high-saturated electron drift velocity. Currently, GaN high power devices are competing well with SiC high power electronics.

The direct band gap of GaN makes it a powerful candidate for many optoelectronics devices as well. A wide array of GaN-based light emitting diodes and lasers are commercially available. One of the advantages GaN holds over other materials is the relatively easy adjustment of the bandgap by incorporating Al (AlGaN). The range of bandgaps achievable is between 3.43 eV (GaN) to 6.2 eV (AlN). This means it can cover the spectral range from 200 to 365 nm, which is highly criti-

Table 6.6 Room temperature mechanical properties of GaN, SiC, and Si [91].

Property	α-GaN (0001)	6H-SiC(0001)	Si (111)
Elastic Modulus (GPa)	330	448	190
Vickers Hardness (GPa)	18	25	3
Fracture Toughness (MPa-$m^{1/2}$)	1.1	3.3	0.7
Thermomechanical Stability	Very good	Very good	Poor

cal for solar-blind UV detectors [89]. Moreover, the ability to make AlGaN/GaN heterostructures allows fabrication of a wide array of electronic devices.

From the MEMS perspective, the piezoelectric property of GaN can be utilized for many transducer applications. Thermal and mechanical properties of GaN are also attractive for harsh environment MEMS applications. Table 6.6 summarizes some of the key mechanical properties of GaN in comparison to SiC and Si. GaN is stiffer and has higher fracture toughness than Si, though SiC is even more rigid. Furthermore, the hardness of GaN remain relatively constant up to 1200 °C and macroscopic dislocation motion and plastic deformation is observed only above 1200 °C [91].

Realization of GaN microsystems requires high purity materials, reasonable fabrication methodology, and device technology for each component of the system. Research has primarily focused on GaN for optoelectronics. Growth of crystalline substrates, deposition of epitaxial films, controlled doping, and metal contacts are at a relatively mature stage, making it possible for device fabrication. These advances in materials stimulate the GaN electronics industry. At its current stage, the GaN electronics industry enjoys a similar status to SiC electronics, and many discrete devices have been developed. GaN high electron mobility transistors (HEMT) are now available on the commercial market. However, from the microsystem prospective, development of integrated circuits is critical, and a research initiative to develop circuits relevant to microsystems is needed.

To date, the primary focus of GaN based development was on electronics and optoelectronics devices; few efforts have been reported on MEMS device fabrication. The limited works reported on GaN based MEMS mainly focus on applying on an AlGaN/GaN heterostructure as a sensing element. The interface properties of an AlGaN/GaN heterostructure is very sensitive to changes in the electrostatic boundary conditions caused by the adsorption of ionic species, interactions with polar liquids, adsorption of gaseous species, and the piezoelectric polarization due to mechanical strain [92]. This phenomenon could be exploited for many transducer applications; however, from the prospective of GaN MEMS, research has to be focused on materials optimization of thin-film mechanical properties for MEMS device fabrication. Furthermore, micromachining strategies need to be developed in order to fabricate free standing functional structures. Some limited work has been reported on fabricating a released GaN MEMS device. One such example is given in Figure 6.17, showing various GaN suspended structures [93].

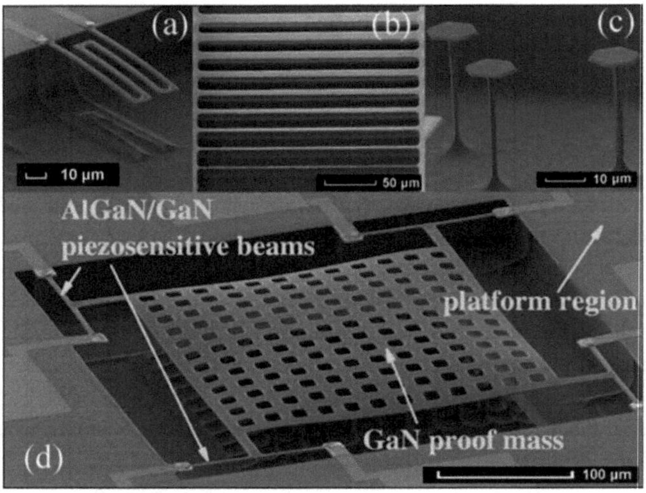

Fig. 6.17 [93] SEM micrograph of various GaN suspended structures. (a) Flexible piezoresistors. (b) Top view of long microbeams. (c) Array of suspended microdisks. (d) MEMS accelerometer featuring a large-area proof mass of 250 by 250 μm^2 with four suspended piezoresistive beams (©IEEE 2009), reprinted with permission.

From the materials prospect, GaN is a well-matched material platform for microsystem applications in demanding environments. The current state of GaN based electronics is extremely encouraging; however, GaN MEMS technology is at an infant stage. Early results on AlGaN/GaN heterostructure sensing elements motivates further research on standalone GaN MEMS as well as GaN integration with SiC MEMS. Thus, GaN should be explored as an alternative or complementing material technology to SiC for harsh environment applications.

6.6 Current state and Future Prospective

In terms of SiC microsystems, the basic components, electronics and MEMS, have shown tremendous progress during the last two decades. Figure 6.18 graphically summarizes the current state of key components of SiC electronics and MEMS in terms of technological maturity. To realize microsystems, an accurate understanding of the current state of each technology at the component level is needed to develop pathways to mitigate shortcomings and remove the roadblocks.

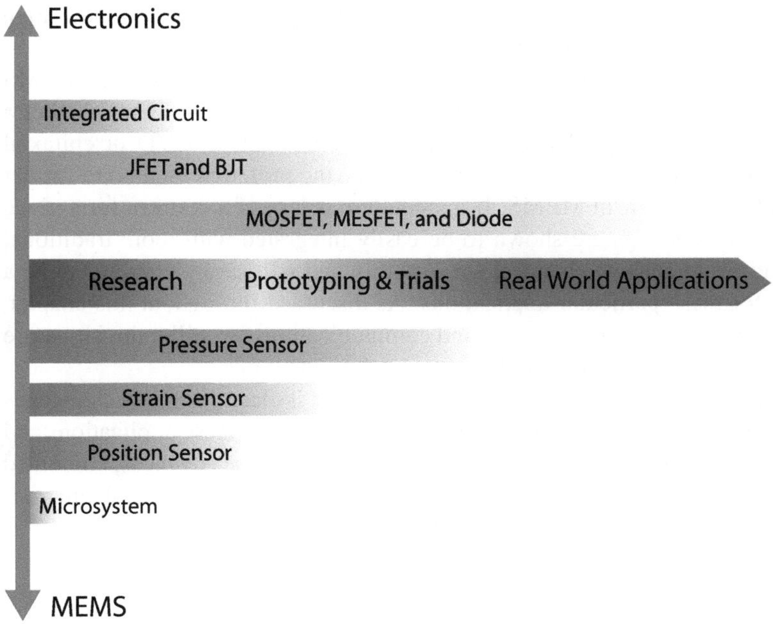

Fig. 6.18 Current state of research and development of SiC devices and systems.

From SiC electronics standpoint, the development of high quality SiC wafers and epitaxial thin films along with the demand for high power and high frequency electronics made SiC based MOSFETs, MESFETs, and Schottky diodes commercial commodities [94, 95]. In parallel, low-resistance ohmic contacts that are stable over thousands of hours above 500 °C have been achieved, allowing fabrication of devices capable of operating at high temperatures [20]. From the perspective of high temperature operation, SiC JFETs are more promising over MOSFETs and MESFETs because they are unipolar devices that do not rely on a high quality semiconductor-dielectric interfaces [96, 97]. Despite the success at individual device level, the integrated SiC circuit aspect is overlooked. From the microsystem aspect, supporting electronic circuits are a critical area that requires immediate attention.

Paralleling electronics development, design, fabrication, and testing of SiC MEMS technology are inching towards the realization of harsh environment compatible sensors and actuators. MEMS grade poly-SiC thin films are routinely deposited in multi-wafer, high-throughput LPCVD reactors [98, 99]. High fidelity micromachining techniques are available with standard etch chemistries and etch masks [100]. These advances in poly-SiC microfabrication technology make it possible to batch fabricate complicated SiC MEMS structures [84]. Some SiC based sensors are ready for real world applications [3]. To push SiC MEMS technology from feasibility to commercial products, long term reliability has to be established. Standardization of

SiC MEMS processes is one particular area that needs focus, which can have high impact on the perception of SiC as viable sensor and actuator platform.

It is clear that the long term path forward is to develop a monolithic MEMS and zero-level encapsulation scheme. However, it is not yet clear as to whether a MEMS-first or IC-first approach, with respective start with LPCVD or epitaxial SiC for the device and scaffold layers, will become the preferred method to further integrate SiC circuitry with MEMS. If temperature is less of a concern for a given application, SiC coatings are shown to be easily integrated with more traditional device materials to achieve chemical survivability and will surely continue to be a useful tool for these particular applications. As discussed throughout this chapter, SiC microsystems, integration, power, and communications are still quite immature and require significant investment of resources.

In order to go from feasibility to useable products, it is clear that many challenges and issues need to be solved. However, SiC is well-suited for these applications and significant technological advances continue to be made. These advances, combined with unprecedented commercial and scientific interest will drive SiC harsh environment microsystems from a dream to a reality.

References

1. Patil A, Fu XA, Neudeck P, Beheim G, Mehregany M, Garverick G (2009). Silicon Carbide Differential Amplifiers for High-Temperature Sensing.Materials Science Forum 600-603:1083–1086
2. Franke AE, Heck JM, King TJ, Howe RT (2003). Polycrystalline Silicon–Germanium Films for Integrated Microsystems. Journal of Microelectromechanical Systems 12(2):160–171
3. Kulite Semiconductor Products Inc.USA. http://www.kulite.com
4. Fedder GK, Howe RT, King Liu T-J, Quévy EP (2007). Technologies for Cofabricating MEMS and Electronics. Proceedings of the IEEE 96(2):306–322
5. Soloviev SI, Gao Y, Sudarshan TS (2000). Doping of 6H-SiC by Selective Diffusion of Boron. Applied Physics Letters 77(24):4004–4006
6. Brand O (2006). Microsensor Integration Into Systems-on-Chip. Proceedings of the IEEE 94(6):1160–1176
7. M. A. Mignardi (1998). BFrom ICś to DMDś,Texas Instruments Technical Jornal 15(3):56–63.
8. BOSCH GMBH, Germany. http://www.bosch.com
9. Freescale Semiconductor Inc. USA.http://www.freescale.com
10. Smith HJ, Montague S, Sniegowski JJ, Murray JR, McWhorter PJ (1995). Embedded Micromechanical Devices for the Monolithic Integration of MEMS with CMOS. Proceedings of 1995 IEDM:609–612
11. Yasaitis J, Judy M, Brosnihan T, Garone P, Pokrovskiy N, Sniderman N, Limb S, Howe RT, Boser B, Palaniapan M, Jiang X, Bhave S (2003). Amodular process for integrating thick polysilicon MEMS devices with sub-micron CMOS. Proceedings of SPIE 4979:145–154
12. Lemkin M, Juneau T, Clar W, Roessig T Brosnihan T (1999). A Low-noise Digital Accelerometer Integrated Using SOI-MEMS Technology.Transducers 99, Sendai, Japan, June 7-10:1292–1297
13. Analog Devices Inc. USA. http://www.analog.com
14. Candler RN, Woo-Tae Park W-T, Li H, Yama G, Partridge A, Lutz M, Kenny T.W (2003). Single wafer encapsulation of MEMS devices. IEEE Transactions on Advanced Packaging 26(3):227–232

15. Messana MW, Graham AB, Yoneoka S, Howe RT, Kenny TW (2010). Packaging of Large Lateral Deflection MEMS Using a Combination of Fusion Bonding and Epitaxial Reactor Sealing. Solid-State Sensor, Actuator and Microsystems Workshop, Hilton Head Island, South Carolina, June 6-10:336–339
16. Bustillo JM, R. T. Howe RT, Muller RS (1998). Surface micromachining for microelectromechanical systems. Proceedings of the IEEE 86(8):1552–1574
17. Infineon Technologies AG. Germany. http://www.infineon.com
18. Fu XA, J. Dunning J, Zorman CA, Mehregany M (2004). Development of a High-Throughput LPCVD Process for Depositing Low Stress Poly-SiC. Materials Science Forum 457-460:305–308
19. Okojie R, Ned A, Kurtz A, Carr W (1996). 6H-SiC pressure sensors for high temperature applications. Proceeding of 9th Annual International Workshop Microelectromechanical Systems, M. Allen and M. Reed, Eds., San Diego, CA, Feb. 1115 1996:146–149
20. Okojie RS, Lukco D, Chen YL, Spry DJ (2002). Reliability assessment of Ti/$TaSi_2$/Pt ohmic contacts on SiC after 1000 h at 600 °C. Journal of Applied Physics 91:6553–6559
21. Kuchuk AV, Guziewicz M, Ratajczak R, Wzorek M, Kladko VP, Piotrowska A (2009). Thermal degradation of Au/Ni2Si/n-SiC ohmic contacts under different conditions. Materials Science and Engineering B 165:38–41
22. Virshup A, Porter LM, Lukco D, Buchholt T K, Hultman L, Spetz AL (2009). Investigation of Thermal Stability and Degradation Mechanisms in Ni-Based Ohmic Contacts to n-Type SiC for High-Temperature Gas Sensors. Journal of Electronic Materials 38 (4):569–573
23. Cook-Chennault KA, Thambi N, Sastry AM (2008). Powering MEMS portable devices—a review of non-regenerative and regenerative power supply systems with special emphasis on piezoelectric energy harvesting systems. Smart Materials and Structures 17:043001
24. Dudney N (2008).Thin Film Micro-Batteries. The Electrochemical Society Interface. Fall 2008: 44–48
25. Excellatron Solid State, LLC. USA. http://www.excellatron.com/high.htm
26. Front Edge Technology, Inc. USA. http://www.frontedgetechnology.com/tech.htm
27. NanoMarkets, LC, USA. http://nanomarkets.net
28. DOE Hydrogen Program. USA. http://www.hydrogen.energy.gov/
29. Dyer CK (2002). Fuel cells for portable applications. Journal of Power Sources 106:31–34
30. MTI MicroFuel Cells. USA. http://mtimicrofuelcells.com/
31. Evans A, Bieberle-Htter A, Rupp JLM, Gauckler LJ (2009). Review on microfabricated micro-solid oxide fuel cell membranes. Journal of Power Sources 194:119–129
32. Bonta PVS, OŃeal CB, Muthusami S (2005). Micro fuel cell technologies, advancements, and challenges. Proceedings of Feulcell 2005:673–682
33. Morse JD (2007). Micro-fuel cell power sources. International Journal of Energy Research 31:576–602
34. Fernandez-Pello AC (2002). Micropower Generation Using Combustion:Issues and Approach. Proceedings of the Combustion Institute (29):883–899
35. Swanger M, Walther DC, Fernandez-Pello AC, Pisano AP (2004). Small-scale rotary engine power system development status. Western States Section / Combustion Institute, Spring 2004, Davis, CA. WSS-04S-9
36. Epstein AH, Senturia SD, Anathasuresh G, Ayon A, Breuer K, Chen KS, Ehrich FE, Gauba G, Ghodssi R, Groshenry C, Jacobson S, Lang JH, Lin CC, Mehra A, Mur Miranda JO, Nagle S, Orr DJ, Piekos E, Schmidt MA, Shirley G, Spearing SM, Tan CS, Tzeng Y-S, Waitz IA (1997). Power MEMS and Microengines. Proceedings of the IEEE Transducers 97 Conference, Chicago, IL, June 1997:753–756
37. Li H, Lal A, Blanchard J, Henderson D (2002). Self-reciprocating radioisotope-powered cantilever. Journal of Applied Physics 92 (2): 1122–1127
38. Guo H Lal A (2003). Nanopower Betavoltaic Microbatteries. Digest of Technical Papers, The 12th International Conference on Solid State Sensors, Actuators and Microsystems, Transducers03:36–39

39. Sims PE, Dinetta LC, Dugancavanagh K, Goetz MA (1995). Gallium Phosphide Energy Converters. Proceedings of the XIV Space Photovoltaic Research and Technology Conference (SPRAT XIV), edited by G. Landis NASA CP-10180:231–236
40. Pool FS, Stella PM, Anspaugh B (1989), GaP betavoltaic cells as a power source. Space Photovoltaic Research and Technology1989, edited by G. Landis NASA CP 3107:359–370
41. Andreev VM, Kevetsky AG, Kaiinovsky VS, Khvostikov VP, Larionov VR, Rumyantsev VD, Shvarts MZ, Yakimova EV, Ustinov VA (2001). Tritium-powered betacells based on AlxGa1-xAs. Conference Record of the Twenty-Eighth IEEE Photovoltaic Specialists Conference-2000:1253–1256
42. Deus S (2001). Tritium-powered betavoltaic cells based on amorphous silicon. Conference Record of the Twenty-Eighth IEEE Photovoltaic Specialists Conference-2000:1246–1249
43. Sun W, Kherani NP, Hirschman KD, Gadeken LL, Fauchet PM (2005). A Three-Dimensional Porous Silicon pn Diode for Betavoltaics and Photovoltaics. Advanced Materials 17(10):1230–1233
44. Shreter YS, Rebane TT, Bochkareva NI (2002). Polymers, Phosphers, and Voltaics for Radioisotope Microbatteries, edited by K. E. Bower, Y.A. Barbanel, Y. G. Shreter, and G. W. Bohnert CRC Press, Boca Raton, FL:365–388
45. Olsen LC (1974). Advanced betavoltaic power sources. Proceedings of the 9th Intersociety Energy Conversion Engineering Conference, edited by Liberman AR, Osmeyer WE, American Society of Mechanical Engineers, New York:754–762
46. Lebedev AA, Kozlovski VV, Strokan NB, Davydov DV, Ivanov AM, Strel'chuk AM, Yakimova R (2002). Radiation Hardness of Wide-Gap Semiconductors (Using the Example of Silicon Carbide). Semiconductors 36(11):1270–1275
47. Chandrashekhar MVS, Duggirala R, Spencer MG, Lal A (2007). 4H SiC betavoltaic powered temperature transducer. APPLIED PHYSICS LETTERS 91:053511
48. Eiting CJ, Krishnamoorthy V, Rodgers S, George T, Robertson JD, Brockman J (2006). Demonstration of a radiation resistant, high efficiency SiC betavoltaic APPLIED PHYSICS LETTERS 88:064101
49. Lee JB, Chen Z, Allen MG, Rohatgi A, Arya R (1995).A miniaturized high-voltage solar cell array as an electrostatic MEMS power supply.Journal of Microelectromechanical Systems 4(3):102–108
50. Bellew CL, Hollar S, Pister KSJ (2003).An SOI process for fabrication of solar cells, transistors and electrostatic actuators. International Conference on Solid-State Sensors, Actuators and Microsystems, Transducers03, Proceedings of the IEEE Transducers 03 Conference, Chicago, IL, June 2003 (2):1075–1078
51. Landis GA (2005). High-Temperature Solar Cell Development. NASA/CP—2005-213431:241–247
52. Ismail BI, Ahmed WH (2009). Thermoelectric Power Generation Using Waste-Heat Energy as an Alternative Green Technology. Recent Patents on Electrical Engineering 2:27–39
53. Rowe DM (2006). Themoelectric Waste Heat Recovery as a Renewable Energy Source. International Journal of Innovations in Energy Systems and Power 1(1):13–23
54. Venkatasubramanian R, Siivola E, Colpitts T, OQ́uinn B (2001). Thin-film thermoelectric devices with high room-temperature figures of merit. Nature 413 (6856):597–602
55. Lee J, M. I. Lei MI, S. Rajgopal S, Mehregany M (2009). Thermoelectric Characterizations of N-Type Polycrystalline Silicon Carbide and Comparison with Conventional Thermopiles. Transducers 2009:1861–1864
56. Xie J, Lee C, Wang M-F, Liu Y Feng H (2009). Characterization of heavily doped polysilicon films for CMOS-MEMS thermoelectric power generators. Journal of Micromechanics and Microengineering 19:125029
57. Wang XH, Yamamoto A, Eguchi K, Obara H, Yoshida T (2003). Thermoelectric properties of SiC thick films deposited by thermal plasma physical vapor deposition. Science and Technology of Advanced Materials:167–172
58. Micropelt GmbH, Germany. http://www.micropelt.com
59. Hi-Z Technology, Inc. USA. http://www.hi-z.com

60. Beeby SP, Tudor MJ, White NM (2006). Energy harvesting vibration sources for microsystems applications. Measurement Science and Technology 17:R175–R195
61. Nye J F (1957). Physical Properties of Crystals. Oxford University Press, First Edition
62. Priya S, Inman DJ (2009). Energy harvesting Technologies, Springer-Verlag, Berlin, Heidelberg, New York
63. Muralt P, Marzenck M, Belgacem B, Calame F, Basrour S (2009). Vibration Energy Harvesting with PZT Micro Device. Procedia Chemistry 1:1191–1194
64. Piazza G, Stephanou PJ, Wijesundara MBJ, Pisano AP (2005). Single-chip multiple-frequency filters based on contour-mode aluminum nitride piezoelectric micromechanical resonators. Solid-State Sensors, Actuators and Microsystems, 2005. Digest of Technical Papers, TRANSDUCERS 05:2065–2068
65. Stephanou PJ, Piazza G, White CD, Wijesundara MBJ, Pisano AP (2007). Piezoelectric aluminum nitride MEMS annular dual contour mode filter. Sensors and Actuators A-Physical A134(1):151–162
66. Avago Technologies. USA. http://www.avagotech.com/
67. Cimalla V, Pezoldt J, Ambacher O (2007). Group III nitride and SiC based MEMS and NEMS: materials properties,technology and applications. Journal of Physsics D: Applied Physics 40:6386–6434
68. Mide Engineering Technologies. USA. http://www.mide.com
69. Beeby SP, Tudor MJ, Torah RN, Roberts S, OĎonnell T, Roy S (2007). Experimental comparison of macro and micro scale electromagnetic vibration powered generators. Microsystem Technologies 13:1647–1653
70. Perpetuum Ltd, UK. http://www.perpetuum.com/
71. Roundy S, Wright PK, Pister KSJ (2002). Micro-electrostatic Vibration to Electricity Converters. ASME International Mechanical Engineering Congress & Exposition, New Orleans, Louisiana 17-22, 2002. IMECE2002-34309
72. Miranda JOM (2004). Electrostatic Vibration to Electric Energy Conversion. PhD Thesis, Massachusetts Institute of Technology, Cambridge MA
73. Roundy S, Wright PK, Rabaey J (2003). A study of low level vibrations as a power source for wireless sensor nodes. Computer Communications 26:1131–1144
74. Roundy S, Leland ES, Baker J, Carleton E, Reilly E, Lai E, Otis B, Jan M. Rabaey JM, Wright PK, Sundararajan V (2005). Improving Power Output for Vibration-Based Energy Scavengers. PERVASIVE computing 4(1):28–36
75. MicroStrain, Inc. USA. http://www.microstrain.com/embed-sense.aspx
76. Von JA, Najafi K (1997). On-Chip Coils With Integrated Cores For Remote Inductive Powering Of Integrated Microsystems. Transducers 97, lnternatimal Coriference on Solid-state Sensors and Actuators Chicago, June 16-19, 1997: 999-1002
77. Suster M, Chaimanonart N, Guo J, Ko WH, Young DJ (2005). Remote-Powered High-Performance Strain Sensing Microsystem. 18th IEEE International Conference on Micro Electro Mechanical Systems MEMS 2005:255-258
78. Arms SW, Townsend CP, Churchill DL, Galbreath JH, Mundell SW (2005). Power Management for Energy Harvesting Wireless Sensors. SPIE Intĺ Symposium on Smart Structures & Smart Materials
79. Smith S, Tang TB, Terry JG, Stevenson JT, Flynn BW, Reekie HM, Murray AF, Gundlach AM, Renshaw D, Dhillon B, Ohtori A, Inoue Y, Walton AJ (2007). Miniaturised Drug Delivery System with Wireless Power Transfer and Communication. IET Nanobiotechnology 1(5):80–86
80. Wang Y, Jia Y, Chen Q, Wang Y (2008). A Passive Wireless Temperature Sensor for Harsh Environment Applications. Sensors 8:7982-7995
81. Neudeck PG, Spry DJ, Chen L-Y, Chang CW, Beheim GM, Okojie RSO, Evans LJ, Meredith RD, Ferrier TL, Krasowski MJ, Prokop NF (2009). Prolonged 500 C Operation of 6H-SiC JFET Integrated Circuitry. Materials Science Forum Vols. 615-617 (2009): 929-932
82. Chen LY, Spry DJ, Neudeck PG, in: Proc. 2006 IMAPS International High Temperature Electronics Conference, Santa Fe, NM, 2006 (International Microelectronics and Packaging Society, Washington, DC, 2006: 240

83. Neudeck PG, Garverick SL, Spry DJ, Chen L-Y, Beheim GM, Krasowsk MJ, Mehregany M (2009). Extreme temperature 6H-SiC JFET integrated circuit technology. Physica Status Solidi A:1–17
84. Myers DR, Cheng KB, Jamshidi B, Azevedo RG, Senesky DG, Chen L, Mehregany M, Wijesundara MBJ, A. P. Pisano AP (2009). Silicon carbide resonant tuning fork for microsensing applications in high-temperature and high G-shock environments. Journal of Micro/Nanolithography. MEMS MOEMS 8(2):021116
85. Brown TG, Davis B, Hepner D, Faust J, Myers C, Muller P, Harkins T (2001). Strap-Down Microelectromechanical (MEMS) Sensors for High-G Munition Applications. IEEE Transactions on Magnetics 37(1):336-342
86. Azevedo RG, Jones DJ, Jog AV, Jamshidi B, Myers DR, Chen L, Fu X-A, Mehregany M, Wijesundara MBJ, Pisano AP (2007). A SiC MEMS Resonant Strain Sensor for Harsh Environment Applications, IEEE Sensors Journal 7(4):568-576
87. Atwell AR, Okokie RS, Kornegay KT, Roberson SL, Beliveau A (2003). Simulation, fabrication and testing of bulk micromachined 6H-SiC high-g piezoresistive accelerometers. Sensors and Acutators A 104:11–18
88. Neudeck PG (2006). Silicon Carbide Technology. The VLSI Handbook, Chapter 5 (Editor Wai-Kai Chen, CRC Press, Second Edition)
89. Duboz J-Y (1999). GaN as Seen by the Industry. Physica status solida (a) 176:5–14
90. Strite S, Morko H (1992). GaN, AIN, and InN: A review. Journal of Vacuum Science and Technology 10(4):1237–1266
91. Yonenaga I (2001).Thermo-mechanical stability of wide-bandgap semiconductors: high temperature hardness of SiC, AlN, GaN, ZnO and ZnSe. Physica B 308–310:1150–1152
92. Pearton SJ, Kang BS, Kim S, Ren F, Gila BP, Abernathy CR, Lin J, Chu SNG (2004). GaN-based diodes and transistors for chemical, gas, biological and pressure sensing, Journal of Physics: Condense Matter 16:R961–R994
93. Lv J, Yang Z, Yan G, Lin W, Cai Y, Zhang B, Chen KJ (2009). Fabrication of Large-Area Suspended MEMS Structures Using GaN-on-Si Platform. IEEE Electron Device Letters 30 (10):1045-1047
94. Cree Inc. USA. http://www.cree.com
95. Hamada K (2009). Present Status and Future Prospects for Electronics in EVs/HEVs and Expectations for Wide Bandgap Semiconductor Devices. Material Science Forum 600-603:889-893
96. Rebello NS, Shoucair FS, Palmou JW (1996). 6H silicon carbide MOSFET- modelling for high temperature analogue integrated circuits (25-500 °C). IEEE Proceedings of Circuits Device Systems 143(2):115–122
97. Spry D, Neudeck P, Okojie R, Chen LY, Beheim G, Meredith R, Mueller W, Ferrier T (2004). Electrical Operation of 6H-SiC MESFET at 500 °C for 500 Hours in Air Ambient. IMAPS International High Temperature Electronics Conference, Santa Fe, NM
98. Zorman CA, Rajgopal S, Fu XA, Jezeski R, Melzak J, Mehregany M (2002). Deposition of Polycrystalline 3C-SiC Films on 100 mm Diameter Si(100) Wafers in a Large-Volume LPCVD Furnace. Electrochemical and Solid-State Letters 5(10):G99–G101
99. Wijesundara MBJ, Valente G, Ashurst WR, Howe RT, Pisano AP, Carraro C, Maboudian R (2004). Single-Source Chemical Vapor Deposition of 3C-SiC Films in a LPCVD Reactor Part I: Growth, Structure, and Chemical Characterization. Journal of the Electrochemical Socciety 151:C210–C214
100. Gao D, Wijesundara MBJ, Carraro C, Howe RT, Maboudian R (2004). Recent progress toward a manufacturable polycrystalline SiC surface micromachining technology. IEEE Senssors Journal 4(4):441–448

Index

accelerometer, 13, 69, 75, 144–146, 148, 220, 224
Acheson, 36, 49
adaptive operation, 10
aerospace, 8, 9, 11, 13, 25, 204, 220
air-to-fuel ratio, 3
aluminum nitride, 98, 175, 195, 210, 211, 222
APCVD, 54, 66–68, 83
automotive, 1, 3–5, 118, 125, 204, 205
avionics, 11

BCB, 171
betavoltaics, 201, 204, 205
bilayer, 15, 46, 47
bipolar junction transistor, 115, 120, 121
bulk-micromachined, 84, 125, 130, 131, 200

carrier concentration, 18–20, 34, 50, 62, 101, 106, 118, 222
chimney reactor, 57
CMOS compatible, 81, 146, 208
combustion control, 1, 4, 5
combustion sensors, 5
condition-based maintenance, 10
contact resistance, 103–105, 174, 191, 199
Coriolis, 148
Curie temp., 213
Curie temperature, 211, 213
CVD, 28, 36, 37, 40–43, 53–55, 57–60, 62–64, 66, 77, 79, 83, 154, 162, 199

Darlington configuration, 115
decomposition, 36
delamination, 28, 200

electrostatic, 84, 159–161, 196, 206, 209, 213–215, 223
environmental power, 201, 206
extensometer, 133

fuel cell, 202–204

gallium arsenide, 18–22
gallium nitride, 207, 222–224
gauge factor, 22–24, 133, 141
gyroscope, 148

hot wall reactor, 56
HTCVD, 40, 63, 66
hybrid integration, 190, 191

interleaved, 192, 194–197, 205
ion-beam assisted deposition, 75–78, 176, 179, 180

junction gate field-effect transistor, 97, 98, 105, 116, 119, 120, 154, 155, 157, 190, 218

lapping, 83, 85
Lely, 36, 37, 49, 63

magnetron sputtering, 66, 78, 154
mechanical polishing, 83, 85
metal semiconductor field-effect transistor, 105, 118, 121
micro engines, 201
micropipes, 49, 50, 52, 64, 65
molding, 80, 83–85, 172, 174
monolithic integration, 28, 67, 148, 185, 190–192, 214

naval vessels, 12
navigation, 8, 9, 14, 144
nucleation, 36, 39, 40, 43, 46–49, 54, 60, 87

ohmic contact, 28, 104, 129, 199, 207

oil, 6–8

photovoltaics, 201, 204, 206
physical vapor transport, 37–45, 47–50, 52, 63
piezoelectric, 9, 133, 209–211, 215, 222, 223
piezoresistor, 22, 130, 131, 133
planarization, 85, 98, 193–195, 198
planetary rotation, 59
polarity, 39, 47, 50, 61, 82, 149
polytype control, 43
power management, 200, 201, 216
power storage, 216
process control, 6, 42

radioactive, 11
real-time monitoring, 3, 6, 9, 11, 132
RF communication, 118, 217
RF powering, 215, 217

scavengers, 201, 206, 211, 214, 216
scavenging, 205, 206, 209–211, 213–216, 222
Schottky contact, 110, 111, 114
Schottky rectifier, 106–109, 111–114
selectivity, 81, 82, 126, 130, 132, 150, 154, 195
SiGe, 172, 193, 207, 208
site competition epitaxy, 59, 61, 62
space exploration, 2, 9, 14, 203
stacking fault energy, 46, 47
step-controlled epitaxy, 59, 60
structural health monitoring, 9–14
sublimation epitaxy, 53
supersaturation, 38, 40, 42, 43, 47, 49, 56, 60
surface-micromachined, 27, 127, 128, 142, 144, 154, 191

temperature sensitivity, 141, 154, 181
thermal expansion, 25, 27, 35, 71, 126, 148, 156, 167, 183
thermistor, 154, 155, 157
thermo-mechanical, 24, 183
thermoelectric, 201, 206–209
thermopile, 208
top seed solution growth, 44, 64
transmission, 121, 122, 200, 201, 217

vibration power, 209, 211, 212, 214, 215, 222

wide-bandgap, 154
wireless communication, 7, 97, 189, 217

ZT, 207, 208

MIX
Papier aus verantwortungsvollen Quellen
Paper from responsible sources
FSC® C105338

If you have any concerns about our products,
you can contact us on
ProductSafety@springernature.com

In case Publisher is established outside the EU,
the EU authorized representative is:
Springer Nature Customer Service Center GmbH
Europaplatz 3, 69115 Heidelberg, Germany

Printed by Libri Plureos GmbH
in Hamburg, Germany